A Color Atlas
of
Histology

A Color Atlas of Histology

Dennis Strete, Ph.D.

McLennan Community College

Consultant, Department of Pathology and Laboratory Medicine

Waco Veterans Affairs Medical Center

HarperCollins*College*Publishers

Executive Editor: Bonnie Roesch
Cover Design: Ellen K. Coolidge
Cover Photos: Dennis Strete
Electronic Production Manager: Eric Jorgensen
Publishing Services and Electronic Page Makeup: Thompson Steele Production Services
Printer and Binder: RR Donnelley & Sons Company
Cover Printer: The Lehigh Press, Inc.

A Color Atlas of Histology

Library of Congress Cataloging-in-Publication Data

Strete, Dennis.
    A color atlas of histology / Dennis Strete. -- 1st ed.
        p.   cm.
    Includes index.
    ISBN 0-673-99190-3
    1. Histology--atlases.   I. Title.
    [DNLM: 1. Histology--atlases.   QS 517 S915c 1995]
QM557. S896 1995
611'. 018 ' 0222--dc20
DNLM/DLC   CIP
for Library of Congress   93-39412

95 96 97 98 9 8 7 6 5 4 3 2 1

DEDICATED TO THE STUDENTS
WHOSE QUEST FOR KNOWLEDGE
NEVER CEASES

CONTENTS

PREFACE

This atlas is a culmination of my interest in histology. It has been designed for use by either undergraduate or graduate students enrolled in allied health, biology, medical, dental, or veterinary programs. It can be used in histology courses or as a supplement to courses in anatomy or anatomy and physiology.

I believe that an understanding of histology requires a general knowledge of the organ systems and their basic morphology and physiology. The first chapter reviews the cell and includes a section pertaining to electron micrographs and light micrographs on chromosome karyotyping. This basic treatment of tissues at the subcellular level is followed by chapters introducing the various types of tissues. Each organ system is then covered. Throughout, I strive to encourage an understanding of the human body on a microscopic level as well as an appreciation of the intricate relationships among the organ systems.

Among the features of this atlas are:

- Over 400 original photomicrographs are included. The best optical microscope and tissue slides available were used for the photography. Magnifications are given for all micrographs.

- Full-color line drawings are included to enhance the understanding of the tissues and their relationships to given organs.

- All micrographs have been carefully labeled. In earlier chapters, simple structures are labeled extensively to give students a better understanding of the subject matter. Subsequent chapters label only the salient structures under discussion.

- Carefully written captions, pointing out key structures and their functions, are included with all figures.

I wish to express my deep appreciation to the many individuals who encouraged me throughout the writing of *A Color Atlas of Histology*. I am indebted to Dr. William D. Hillis, Baylor University, for his generosity in providing medical histology slides; to Dr. Joyce Wilkinson, Scott and White Medical Center, for providing some of the electron micrographs; to Dr. Nicholas Newman, for his suggestions on the lymphatic and cardiovascular systems; and to Dr. Keith Young, Waco Veterans Affairs Medical Center, for his suggestions on the nervous system. My very special thanks go to Dr. Charles Conley, pathologist, for his support and encouragement, and for the use of his pathology laboratory at the Waco Veterans Affairs Medical Center where some of the tissue slides were photographed.

My sincere thanks and appreciation go to Lorraine Stansel, editing consultant, for her review of the first draft of the manuscript; to Kathy Johnson, transcriptionist for the project; to Charles Hickman, HarperCollins*CollegePublishers* senior sales representative, who was instrumental in establishing the link between the author and the publisher; to Bonnie Roesch, executive editor at HarperCollins*CollegePublishers*; to Craig Kirkpatrick, copyeditor; and to Thompson Steele Production Services. My gratitude and thanks go to HarperCollins*CollegePublishers* for its financial support and its confidence in my ability to pursue this project.

My deepest appreciation and thanks go to Dr. B. J. Martin, my mentor, who rekindled my interest in histology, and to Dr. Billy Ballard, pathologist, for encouraging me to explore other avenues in academia.

I sincerely hope that you and your students will find this atlas beneficial to your course of study, and that knowledge gained from its use will aid your students in their future endeavors. I welcome your comments and suggestions.

Dennis Strete, Ph.D.

The Cell

The **cell** forms the basic structural and functional unit of a tissue. The **tissues,** in turn, form four morphological and functional divisions of the body: **epithelial tissue, connective tissue, nervous tissue,** and **muscular tissue.** Every mature, living cell contains a **nucleus, cytoplasm, organelles, inclusions,** and a highly specialized limiting membrane called the **plasma membrane** or **plasmalemma.** The nucleus stores the genetic information and coordinates all organelle and cellular metabolic activities.

Cell Membrane (plasmalemma) The plasmalemma is a trilaminar membrane structure (a lipid layer between two protein layers) that measures approximately 8 to 10 nm in thickness. The external protein layer of the membrane may be lined by polysaccharide units that form the outer uneven glycocalyx coating. The thickness of the glycocalyx coating may vary depending on the type and function of the cell.

Protoplasm The protoplasm displays physiological characteristics such as conductivity, contractibility, irritability, absorption, secretion, excretion, and metabolism. These characteristics of the protoplasm are important if the cell is to survive, divide, and engender cell-to-cell communication.

Cell Organelles Cytoplasmic organelles are specific functional units of the cell. Included in the organelles are such diverse structures as agranular and granular endoplasmic reticula, plasmalemmas, golgi complexes, ribosomes, cilia, lysosomes, peroxisomes, mitochondria, centrioles, filaments, and microtubules. Most of the organelles are surrounded by a limiting membrane that is morphologically similar to the plasma membrane. Intermediate filaments in the cell matrix can be of many types (nonfibrillary, desmoid, keratin, acidic, and protein) depending on the function of the cell. Other cytoplasmic components of the cells may include pigments, glycogen and lipid molecules, vacuoles, and undifferentiated granules.

The Mammalian Cell

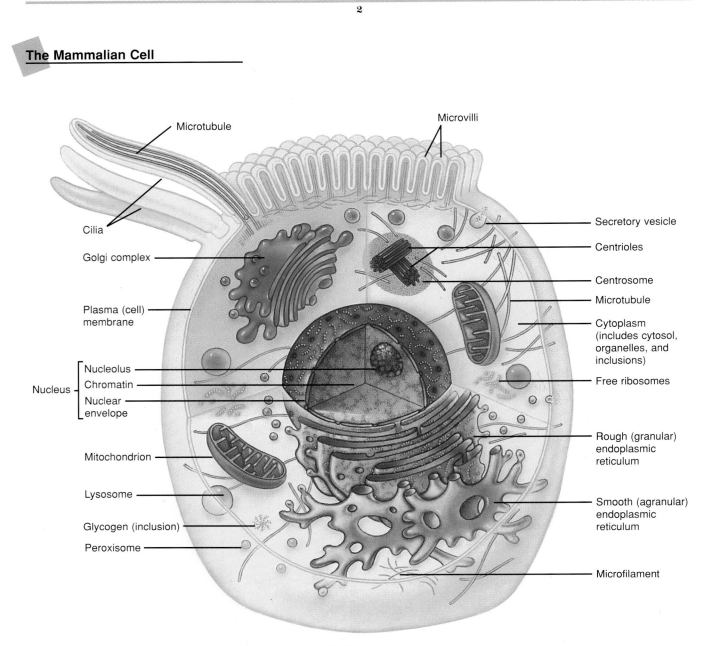

Microtubule

Microvilli

Cilia

Golgi complex

Plasma (cell) membrane

Nucleus
  Nucleolus
  Chromatin
  Nuclear envelope

Mitochondrion

Lysosome

Glycogen (inclusion)

Peroxisome

Secretory vesicle

Centrioles

Centrosome

Microtubule

Cytoplasm (includes cytosol, organelles, and inclusions)

Free ribosomes

Rough (granular) endoplasmic reticulum

Smooth (agranular) endoplasmic reticulum

Microfilament

Sectional view

FIGURE 1.1
Generalized diagrammatic representation of an animal cell based on electron microscopic studies. Not all structures shown in this diagram—e.g., the cilia, microtubules, and microvilli—are present in all living cells. The structures in the diagram have been drawn to facilitate the understanding of a living cell and its function.

FIGURE 1.2
Transmission electron micrograph (TEM) of a freeze fractured cell. The nucleus reveals pores in the membrane. The elements of the rough endoplasmic reticulum and golgi complex can be resolved in this micrograph. (30,000×)

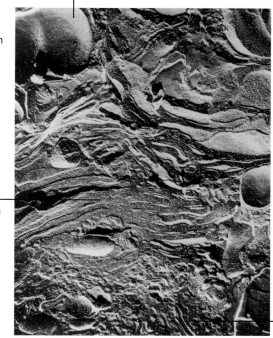

Nucleus

Rough endoplasmic reticulum

Golgic complex

Nucleus and Nucleolus

FIGURE 1.3
TEM of a cell nucleus exhibiting the characteristic appearance of a nuclear membrane, granular heterochromatin, dispersed euchromatin, and a nucleolus. The nucleolus shows distinct areas of pars granulosa and pars fibrosa. A nuclear envelope surrounds the nucleus. (6000×)

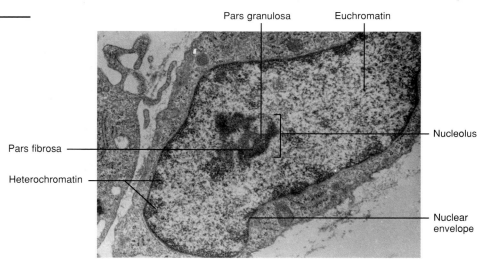

Pars granulosa

Euchromatin

Pars fibrosa

Nucleolus

Heterochromatin

Nuclear envelope

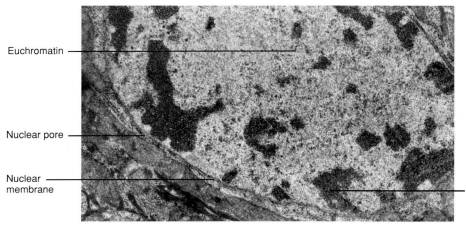

Euchromatin

Nuclear pore

Nuclear membrane

Heterochromatin

FIGURE 1.4
TEM of a nucleus depicting the nuclear membrane and its bilamellar arrangement. Nuclear pores, which can be observed at intervals, form the main communicating channels between the nucleus and the surrounding cytosol. As in the preceding micrograph, concentrations of euchromatin and heterochromatin can be seen throughout the nuclear matrix. (10,000×)

Cell Organelles

FIGURE 1.5
TEM of a mitochondrion, showing two distinct membranes, an inner and an outer membrane. The membranes are separated by a small interlamellar space. The inner mitochondrial membrane forms numerous folds called cristae. Intracristalar spaces form narrow areas between cristae. Granules can be observed within the matrix. (12,000×)

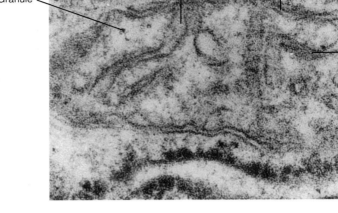

Intracristalar space
Interlamellar space
Granule
Cristae

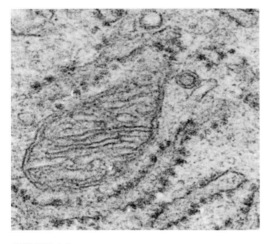

FIGURE 1.6
TEM of a mitochondrion surrounded by rough endoplasmic reticulum. As in the preceding micrograph, the mitochondrion shows the distinct bilamellar arrangement of the inner and outer mitochondrial membranes (10,000×)

Mitochondria

FIGURE 1.7
TEM of concentration of mitochondria as seen in an active epithelial cell. Below the mitochondria, the bilamellar nuclear membranes and nuclear matrix can be observed (7500×).

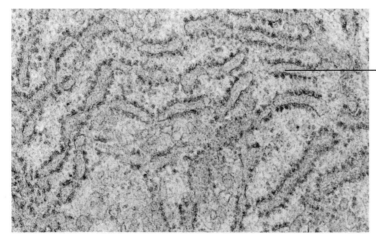

Rough endoplasmic reticulum

FIGURE 1.8
TEM of cytosol showing parallel rows of rough endoplasmic reticulum cisternae. Ribosomes line the membranes of the rough endoplasmic reticulum. (10,000×)

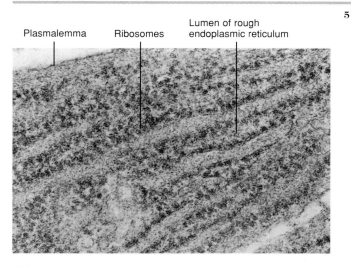

Plasmalemma   Ribosomes   Lumen of rough
endoplasmic reticulum

FIGURE 1.9
TEM of a fibroblast cell with dense concentrations of ribosomes. A distinct lumen
of the rough endoplasmic reticulum can
be seen between the cisternal membranes. The plasmalemma of the cell surrounds the fibroblast. (12,000×)

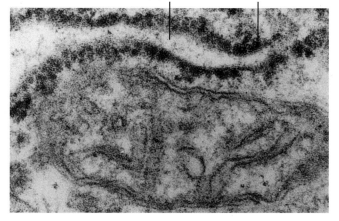

Intralamellar   Rough endoplasmic
space   reticulum

FIGURE 1.10
High-magnification TEM of an endoplasmic reticulum, showing
ribosomes lining the membrane, producing a rough appearance.
The intralamellar space of the endoplasmic reticulum forms the
lumen for the transport of proteins and enzymes. (18,700×)

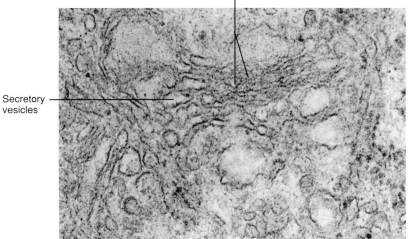

Golgi saccules

Secretory
vesicles

FIGURE 1.11
TEM of a golgi complex (golgi apparatus) with five
flat lamellar cisternae (golgi saccules). The cisternae form disc-shaped structures compressed in
the middle but lobulated peripherally. The lobulated
ends are separate from the main complex and form
secretory vesicles. (70,000×)

FIGURE 1.12
Light micrograph (LM) of neurons. A large number
of golgi bodies can be seen throughout the neuro-
plasm of the neurons. The golgi complexes lie
between the nucleus and the limiting membrane of
the cell. (1000×)

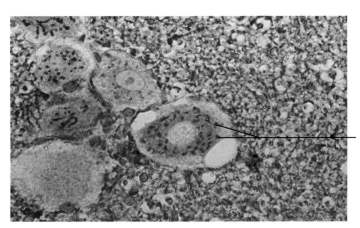

Golgi bodies

Modified Cells And Their Structures

FIGURE 1.13
TEM of two adjacent cells, showing plasmalemmas, intercellular space, and matrix. The lower cell shows concentration of heterochromatin and euchromatin. A nuclear pore disrupts the continuity of the nuclear membrane. In the upper cell, a mitochondrion forms a tubular structure. (12,000×)

Mitochondrion   Heterochromatin

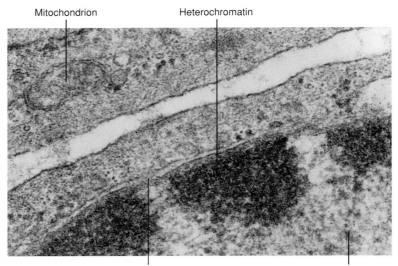

Nucleopore   Euchromatin

Rough endoplasmic reticulum   Plasmalamellar extensions

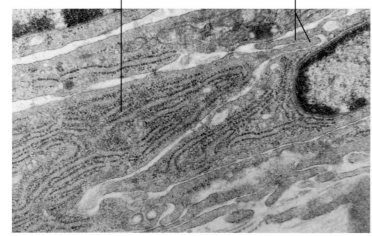

FIGURE 1.14
TEM of fibroblast cells and their irregular plasmalamellar extensions, which may be precursors of collagen fibers. Cells are crowded with rough endoplasmic reticulum. Collagen fibers can be seen in the lower portion of the micrograph. (6000×)

FIGURE 1.15
TEM of a macrophage cell (histiocyte). The cell membrane is irregular and has blunt processes. Distinct concentrations of phagosomes and residual bodies are visible. The structure on the left is the nucleus of the cell. (5,000×)

Heterophagic vacuoles

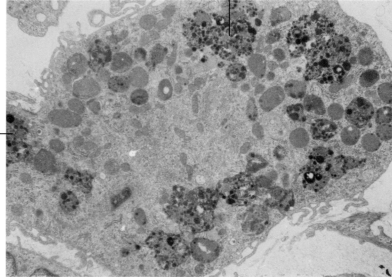

Nucleus

Lysosome

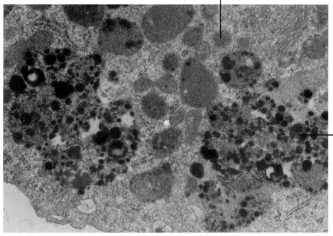

Phagosome

FIGURE 1.16
TEM of heterophagic vacuole (phagosome) as seen in a
macrophage cell. The heterophagic vacuole is formed by the fusion
of a phagosome and a lysosome. (10,000×)

Interlocking structures

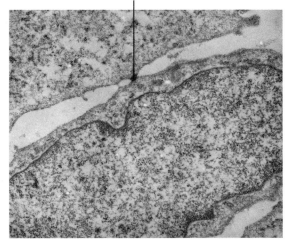

FIGURE 1.17
TEM of epithelium, showing a nexus or gap junction in
which the membranes of adjacent cells form interlocking
structures. Gap junctions are believed to form cell-to-cell
adhesions that facilitate intercellular transport. (30,000×)

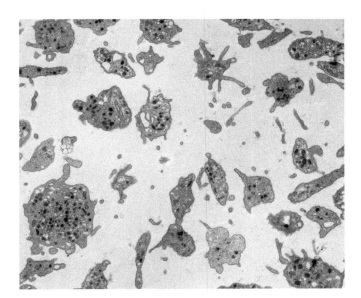

FIGURE 1.18
TEM of platelets (thrombocytes), which are anucleated bodies of
large megakaryocyte cells found in the bone marrow. Platelets
have granular granulomeres or chromomeres surrounded by
pale, granular-free hyalomeres. The granulomeres contain lysosomes, tubules, fibrils, and vesicles. Hyalomeres form the cytoplasmic matrix of the cell. (15,000X)

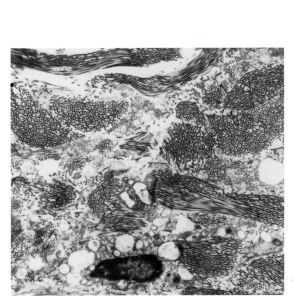

FIGURE 1.19
TEM of collagen fibers in the matrix of a cell. The
fibers follow an irregular course and vary in thickness
(1–10 nm) and length (approximately 300 nm).
Collagen fibers consist of separate parallel fibrils with
an independent collagen unit. (5000×)

FIGURE 1.20
TEM of collagen fibers, showing unit fibrils. The fibrils exhibit repeating bands of fibrils spaced at 640-nm intervals. The repeating bands are tropocollagen molecules measuring 200 nm in length and 1.5 nm in diameter. These molecules coalesce to form collagen fibers. (10,000×)

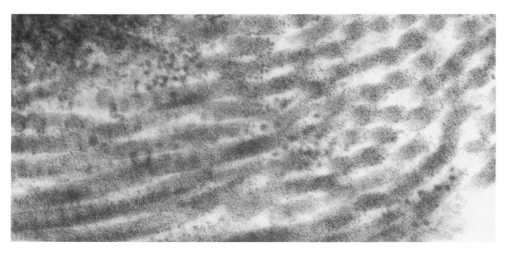

FIGURE 1.21
TEM of cilia, which are motile appendages of specialized cells. The membrane of a cilium is continuous with the cell membrane. A cilium contains 20 bands of protein microtubules, which form nine doublets arranged in a circular manner at the periphery. A pair of microtubules forms the central core of the cilium. (90,000×)

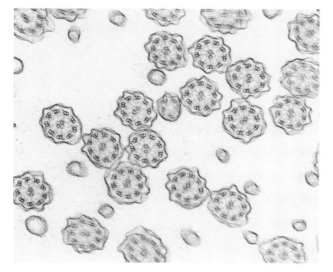

FIGURE 1.22
TEM of the intracellular pigment melanin, as seen in lower epithelial cells called melanocytes. Melanin is enclosed in membrane-bound sacs known as melanosomes. Melanin is also found in the iris and retina of the eye and in some brain cells. The nucleus is eccentrically placed in this micrograph. (35,000×)

Nucleus

Melanasomes

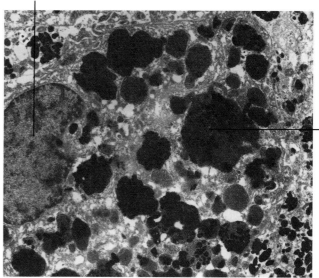

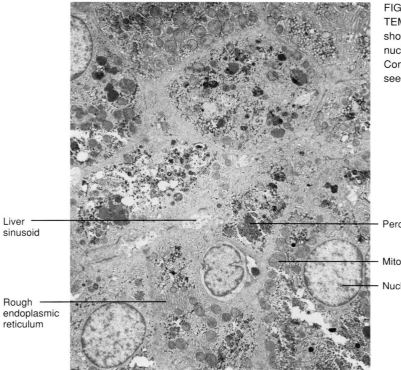

FIGURE 1.23
TEM of hepatocytes of the liver. The cells distinctly show rough endoplasmic reticulum, mitochondria, nuclei, glycogen granules, and liver sinusoids. Concentrations of peroxisomes or microbodies can be seen in the matrix of the cell. (6000×)

Liver sinusoid

Peroxisome

Mitochondria

Nucleus

Rough endoplasmic reticulum

FIGURE 1.24
TEM of a blood vessel. The lumen is lined by endothelial cells and a few elastic fibers, which form the tunica intima. The tunica media of the vessel shows smooth muscle cells surrounded by collagen fibers. The tunica externa is lined by collagen fibers which are visible in both longitudinal and cross-sectional views. (10,500×)

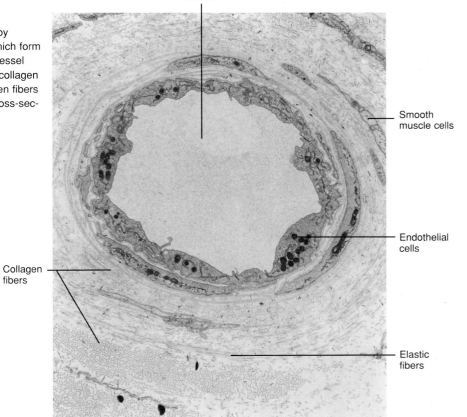

Lumen

Smooth muscle cells

Endothelial cells

Collagen fibers

Elastic fibers

Smooth
muscle cells

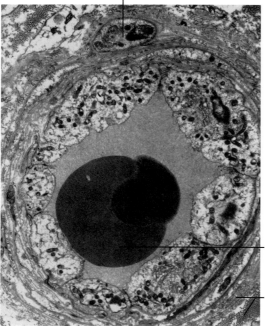

Erythrocyte

Collagen
fibers

FIGURE 1.25
TEM of a blood vessel. The lumen, which is lined by
endothelial cells, is partially occluded by an erythrocyte. The tunica media of the vessel shows smooth
muscle cells and collagen fibers. (18,500×)

Virus

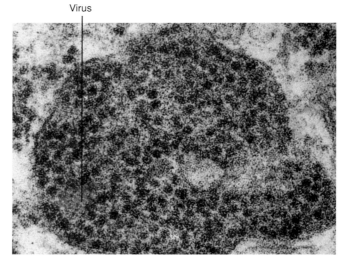

FIGURE 1.26
TEM of a mitochondrion infiltrated with virus. The mitochondrion is
disintegrating, releasing the virus into the cytosol. (150,000×)

Smooth
endoplasmic
reticulum

Golgi
complex

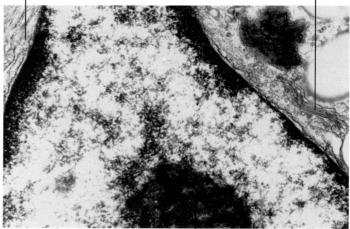

Nucleolus

FIGURE 1.27
TEM of smooth endoplasmic reticulum characterized by the
absence of ribosomes. In this micrograph, the golgi complex and
nucleolus are distinctly clear. The smooth endoplasmic reticulum
synthesizes lipids and is responsible for intracellular transport.
(45,000×)

Nucleus   Centrosome

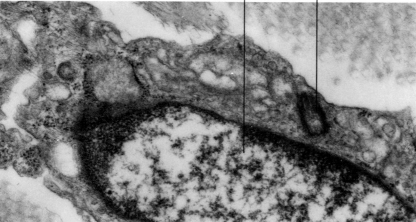

FIGURE 1.28
TEM of centrosome, or the cytoskeleton orga-
nizing center, generally located near the nucleus.
Within the centrosome is a pair of centrioles that
consist of nine triplets of microtubules arranged
in a circular manner near the periphery of the
centriole. (12,000×)

Centriole–centrosome complex

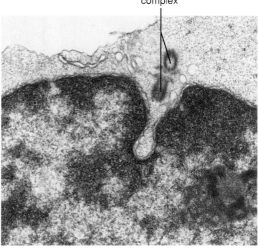

FIGURE 1.29
TEM of a centriole-centrosome complex. The centrioles are at right angles to each other. The microtubular arrangement cannot be discerned at this magnification. (51,000×)

Golgi apparatus

Collagen fibers

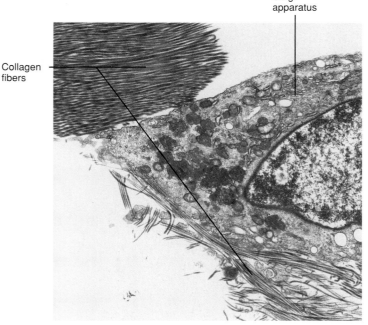

FIGURE 1.30
TEM of a concentration of collagen fibers in epithelial (skin) tissue. Note the golgi apparatus within the cell. Large golgi vesicles are visible near the golgi complex. (18,000×)

Mitochondria

Gap junction

Phagosomes

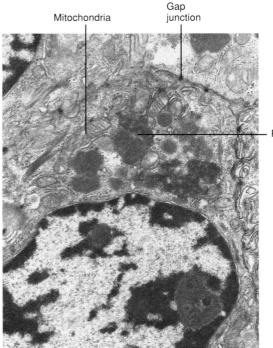

FIGURE 1.31
TEM of a nucleus showing distinct nuclear pores at intervals throughout the nuclear membrane. The micrograph also shows several gap junctions between cells. Mitochondria and phagosomes can also be seen in the cytoplasm of the cell. (30,000×)

Nucleus

Lysosome

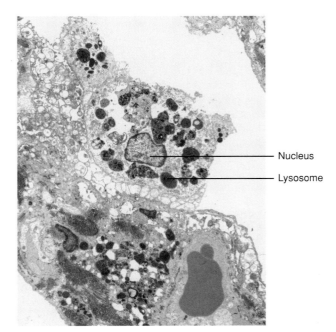

FIGURE 1.32
TEM of an alveolar macrophage with irregularly shaped plasma membrane and blunt cytoplasmic processes. The nucleus is indented, a characteristic feature of the macrophage. The micrograph also shows large numbers of lysosomes in the cytoplasm. The large cell below is an endothelial cell of the blood capillary. (6000×)

Fat vesicles

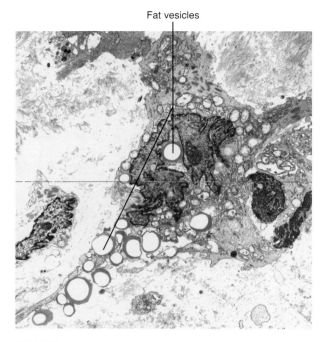

FIGURE 1.33
TEM of a mesothelial cell with large numbers of fat vesicles. The vesicles are composed of glycerol esters and fatty acids that serve as energy storage for the cell. (7500×)

Lipid granules

Glycogen

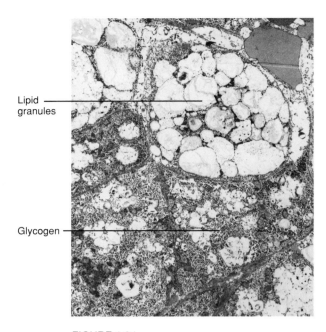

FIGURE 1.34
TEM of hepatocytes (liver cells). The hepatocytes are large polyhedral-shaped cells. Large concentrations of glycogen and lipid granules can be seen throughout the cytoplasm of the cell. (6000×)

FIGURE 1.35
LM of human chromosomes isolated from *in vitro* cell culture at the onset of mitosis. Cross-banding along the lengths of the chromosomes is revealed by enzymatically treating chromosomes with trypsin.

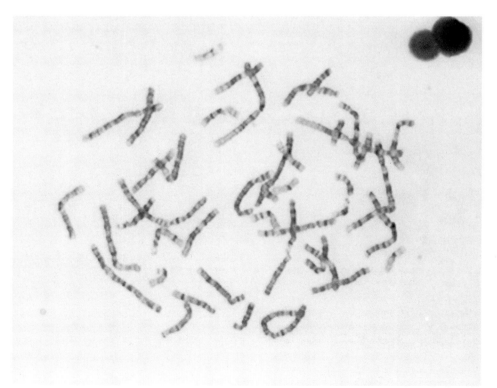

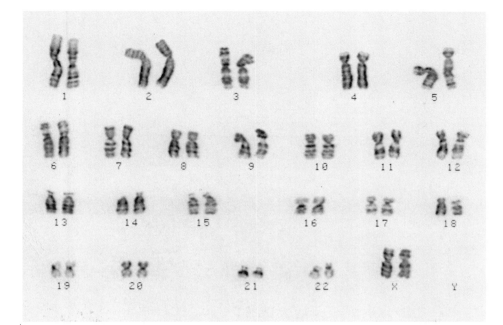

FIGURE 1.36
LM illustrating the karyotype of human female chromosomes arrested in the metaphase stage of mitosis, *in vitro*, by colchicine treatment. The chromosomes were enzymatically treated with trypsin to reveal a cross-banding pattern. In mitotic division, each chromosome consists of a duplicate chromosome, and members of a duplicate chromosome are referred to as an homologous pair. The paired chromatids are joined at a point called the centromere. The chromosomes in the micrograph are arranged to show 22 pairs of autosomes (1–22), and a pair of sex chromosomes, XX. The absence of a Y chromosome indicates that this is a karyotype of a female. On the basis of their size and shape, the chromosomes are classified into seven groups. Group A includes chromosome pairs 1–3; Group B, pairs 4 and 5; Group C, pairs 6–12; Group D, pairs 13–15; Group E, pairs 16–18; Group F, pairs 19 and 20; and Group G, pairs 21 and 22. Females possess 23 pairs of homologous chromosomes since they have two X chromosomes.

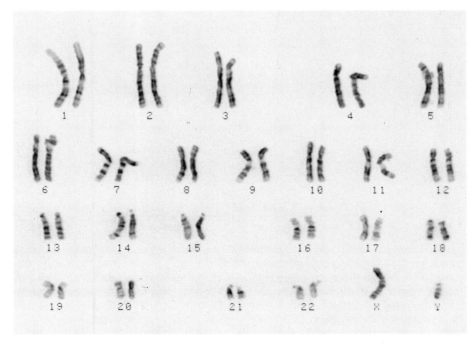

FIGURE 1.37
LM showing the karyotype of human male metaphase chromosomes, prepared from white blood cells cultured *in vitro*. Each chromosome pair is identified by its characteristic banding pattern. As in the preceding micrograph of the female karyotype, the chromosomes are arranged in seven groups. In human males, there are 22 pairs of homologous chromosomes and a pair of sex chromosome, XY.

Mitosis (Whitefish)

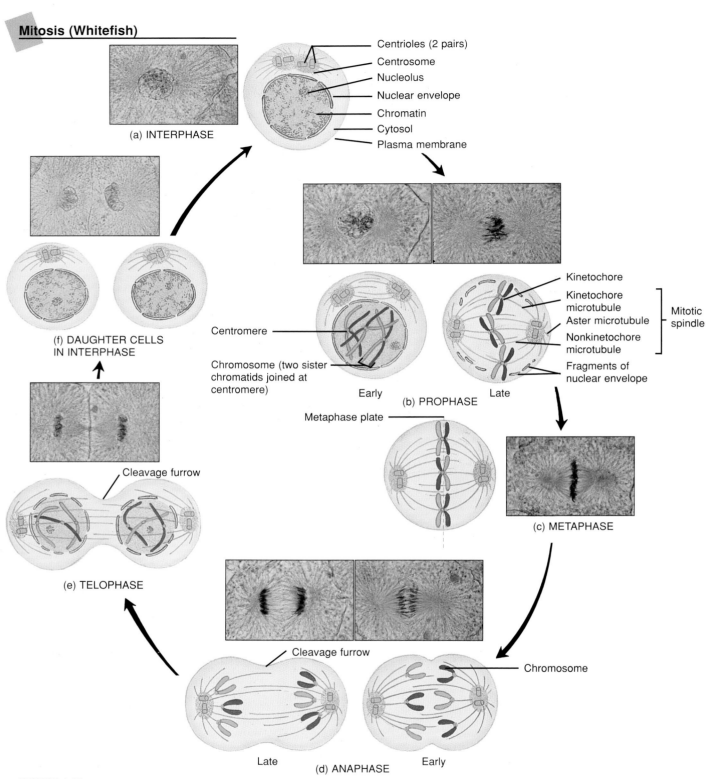

(a) INTERPHASE

Centrioles (2 pairs)
Centrosome
Nucleolus
Nuclear envelope
Chromatin
Cytosol
Plasma membrane

(f) DAUGHTER CELLS IN INTERPHASE

Centromere

Chromosome (two sister chromatids joined at centromere)

Early

Kinetochore
Kinetochore microtubule
Aster microtubule
Nonkinetochore microtubule

Mitotic spindle

Fragments of nuclear envelope

Late

(b) PROPHASE

Metaphase plate

(c) METAPHASE

Cleavage furrow

(e) TELOPHASE

Cleavage furrow

Chromosome

Late

(d) ANAPHASE

Early

FIGURE 1.38

Diagrammatic and photographic representation of cell division—mitosis and cytokinesis—as seen in animal cells. (See also Figure 1.39.) In diagram (a), the cell is in the late interphase stage of the cell cycle. Diagram (b) shows the cell in the early and late prophase stages; the chromatin has condensed into chromosomes. In diagram (c), the cell is in the metaphase stage; the chromosomes have replicated themselves, and are aligned in the center of the cell. In diagram (d), the cell is in the anaphase stage; the chromosomes are separating and moving to their respective poles. In diagram (e), the cell cytoplasm is going through cytokinesis (cytoplasmic division), the nuclear membrane and organelles are reappearing, and the chromosomes are disorganizing into chromatin. Diagram (f) shows the formation of new daughter cells.

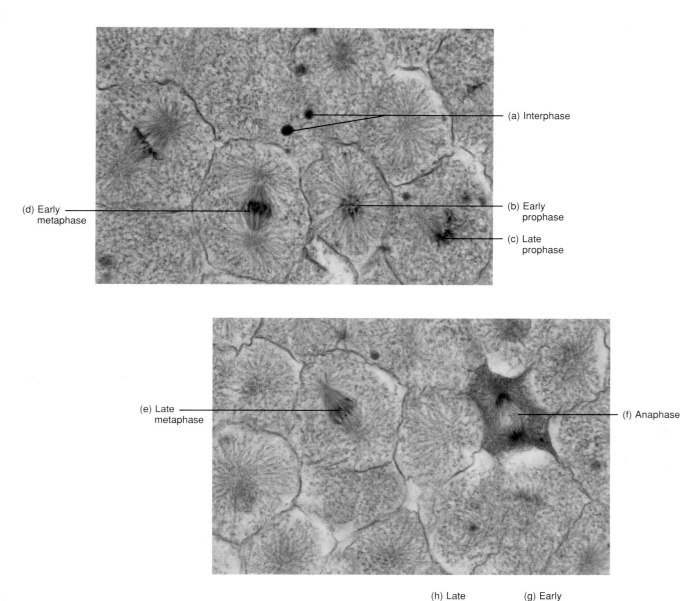

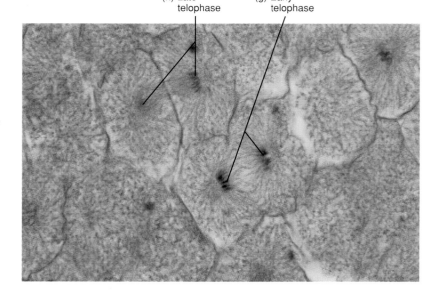

FIGURE 1.39

Three light micrographs (LM) showing cells at various stages of the cell-division cycle illustrated diagrammatically in Figure 1.38. (a) Nuclei of two cells at the interphase stage; these nuclei have not yet differentiated into chromosomes. (b) In the early prophase stage, the chromatin coalesces to form chromosomes. (c) In the late prophase, the chromosomes become aligned in the center of the cell. (d) In the early metaphase, the chromosomes are arranged in a central plane. (e) In the late metaphase, the chromosomes begin to split. (f) In the anaphase stage, the chromatids move toward the opposite poles. (g) In the early telophase, the mitotic spindle disintegrates, and the plasma membrane constricts in the center to separate the cytoplasm into two halves (cytokinesis). (h) In the late telophase, a horizontal cell membrane separates the two daughter cells. (All at 1000×)

Epithelium

There are four basic types of body tissues: **epithelial, connective, muscular, and nervous.** These tissues are found in all organs of the body and function in close association with each other. Epithelial tissue covers free surfaces, lines body cavities and tubular structures, assumes many secretory and lubrication functions, and protects organs from invasion by microorganisms. Modified epithelial tissue structures play an important role in the coordination of muscle, bone, nervous, digestive, reproductive, respiratory, and other systems of the body.

Epithelial tissue may invaginate in certain organs to form exocrine glands. In some cases, epithelial cells take on a completely different role, forming endocrine glands which secrete their products into the vascular system. The epithelial tissue classification is based on the shape of the cells (squamous, cuboidal, or columnar), layering of cells (simple or stratified), and modification of surface

cells (cilia, microvilli, stereocilia, keratinized or nonkeratinized).

As stated earlier, the classification of epithelial tissue is based on the shape and arrangement of cells that overlie a **basement lamina** or membrane. The lamina also separates the epithelial tissue from the underlying connective tissue. A group of epithelial cells that forms a single layer of flat, irregular cells on the lamina is known as **simple squamous epithelium.** In the lymphatic and cardiovascular systems this layer is the endothelium. In the lining of the peritoneal, pleural, and pericardial cavities, the epithelium is called mesothelium. Owing to the thin, flat shape of simple squamous cells, passive transport of fluid metabolites, nutrients, and gases can readily take place between the vascular system and the epithelial cells.

Simple columnar epithelial cells have greater height than width. They are often found in the gastrointestinal tract, where they function as secretory and absorptive cells. They are also found in the bronchi, and as secretory cells in the oviduct and uterus.

The **simple cuboidal cells** are cubic in shape. They function as secretory or absorptive cells (kidney). They also line the ducts of glandular tissue.

The **stratified squamous epithelium** forms several cell layers in the skin, oral cavity, pharynx, esophagus, anal canal, and vagina. The epithelium forms a protective covering with the capacity to replace dead cells periodically. The stratified squamous epithelium may be keratinized or nonkeratinized. Nonkeratinized epithelium lines most cavities (mouth, pharynx, esophagus, vagina, and anal canal).

Keratinized epithelium forms the upper layer of the skin and the gingiva lining of the teeth. **Stratified cuboidal epithelium** and **stratified columnar epithelium** are not common. They are generally found in the ducts of the pancreas, the sweat glands, and the salivary glands. The stratification of cells does not exceed more than two or three layers.

Transitional epithelium changes shape when stretched. The surface cells take on a dome-shaped appearance when relaxed and a squamous shape when stretched. Transitional epithelium lines the urinary bladder and the urinary tract.

Simple Squamous Epithelium

FIGURE 2.1
Light micrograph (LM) of a Bowman's capsule located at the vascular pole of the renal corpuscle. The outer (parietal) and inner (visceral) layers of the corpuscle are lined with simple squamous epithelium. The glomerular capillaries are also lined with simple squamous epithelium. This epithelium is called endothelium. (400×)

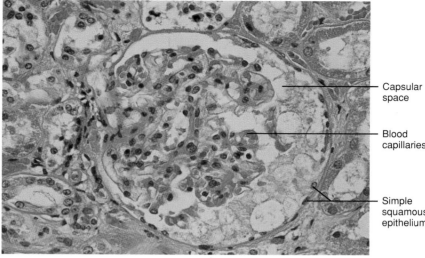

— Capsular space

— Blood capillaries

— Simple squamous epithelium

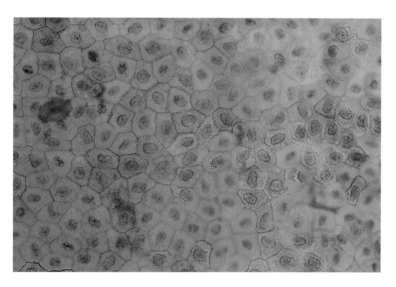

FIGURE 2.2
LM of simple squamous epithelial cells in a surface view of mesothelium. Note the centrally placed nucleus. The mesothelium lines pericardial, peritoneal, and pleural cavities. (400×)

FIGURE 2.3
LM of a blood vessel in cross section. The lumen of the blood vessel is lined with simple squamous epithelium. Here, as in the blood capillaries, the epithelium is known as endothelium. (400×)

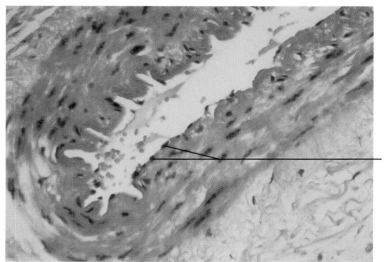

— Endothelium

FIGURE 2.4
LM of simple squamous cells lining a
lymph vessel. The epithelial cell lining in
this case is also known as endothelium.
(200×)

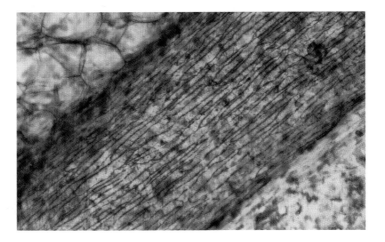

Simple Cuboidal Epithelium

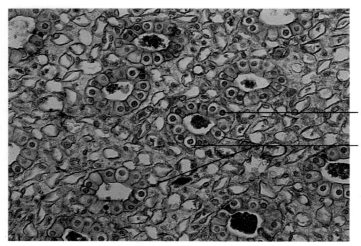

FIGURE 2.5
LM of simple cuboidal epithelium in a collecting tubule of the
kidney medulla. The cuboidal cells represent an intermediate
form between simple columnar and simple squamous cells. The
epithelial cells are cuboidal in shape and have large, centrally
placed nuclei. (400×)

Nucleus

Cuboidal
cells

Cuboidal
cell

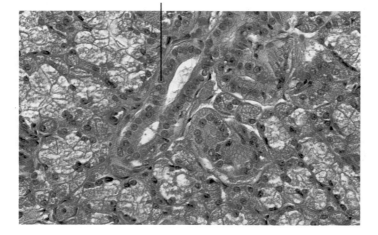

Cuboidal
cells

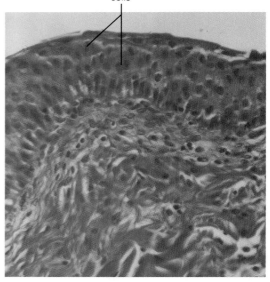

FIGURE 2.6
LM of simple cuboidal epithelium lining a secretory duct in a submaxillary gland. (400×)

FIGURE 2.7
LM of simple cuboidal epithelium in the outermost cellular layer of an ovary. Note the typical cuboidal cells
with an underlying basement lamina. (400×)

Cuboidal
cell

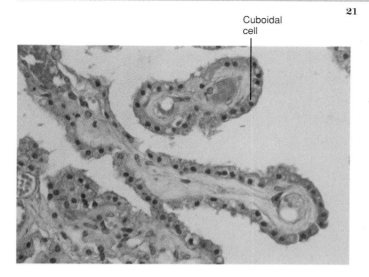

Cuboidal
cell   Colloid

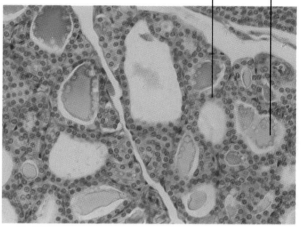

FIGURE 2.8

LM of simple cuboidal epithelium in the outermost covering of choroid plexus located in the ventricles of the brain. The plexus is responsible for secreting cerebrospinal fluid. (400×)

FIGURE 2.9

LM of thyroid follicles lined by simple cuboidal epithelium. Follicular cells secrete a glycoprotein colloidal substance known as thyroxine. Some of the follicular lumina are filled with the colloidal substance. The lightly stained cells between the follicles are parafollicular cells that secrete calcitonin. (400×)

Simple Columnar Epithelium

FIGURE 2.10

LM of simple columnar epithelium lining of the gall-bladder. The function of the columnar cells is to absorb water and concentrate bile. The luminal plasma membrane surface of the cell is modified to form small microvillus-like projections that increase the absorptive surface area of the cell. The presence of microvilli create the striated brush border appearance of the columnar cells. (400×)

Brush   Columnar
border   cells

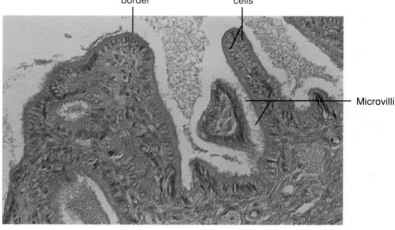

Microvilli

Goblet
Mucus   Lumen   cell

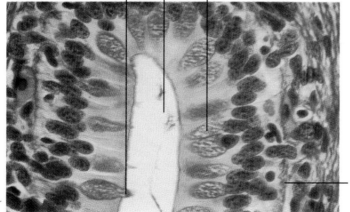

FIGURE 2.11

LM of simple columnar epithelium as seen in the small intestine (jejunum). Some of the cells are modified and form goblet cells, which are mucus-secreting unicellular glands. In this micrograph, a distinct basal lamina is visible below the cells. Some of the goblet cells are actively secreting mucus into the lumen of the gastrointestinal tract. (200×)

Basement
membrane

Columnar cells

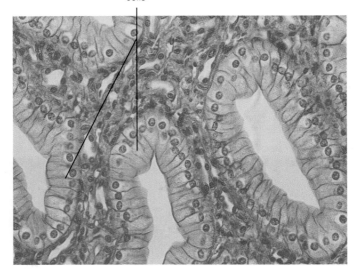

Columnar cells

FIGURE 2.12
LM of simple columnar cells lining the collecting ducts in the medulla of the kidney. Collecting tubules combine to form the larger collecting ducts. The collecting ducts are identified by the presence of tall columnar cells and have lumina larger than those of the collecting tubules. (400×)

FIGURE 2.13
LM of simple columnar cells lining the excretory ducts in a submandibular gland. Not all ducts in the gland are lined by columnar cells. Some of the smaller excretory ducts are lined by simple cuboidal epithelium. (400×)

Cilia   Nucleus   Villus

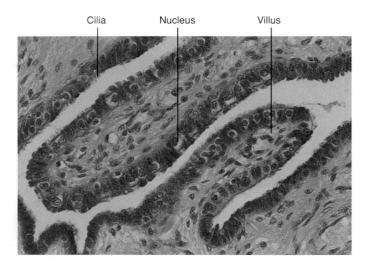

FIGURE 2.14
LM of simple ciliated columnar epithelium in an oviduct (fallopian tube). The inner lining of the oviduct is highly folded, forming villus-like extensions into the lumen. Cilia line the outer border of the columnar cells and, by their rhythmic movement, facilitate the transport of ova to the uterus. (400×)

Nucleus   Cilia

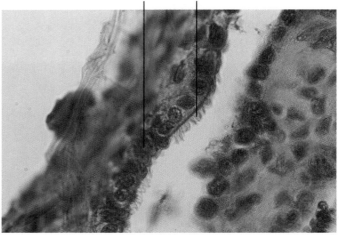

FIGURE 2.15
LM of simple ciliated columnar epithelium of the oviduct magnified to delineate tufts of cilia bordering the columnar cells. (1000×)

FIGURE 2.16
LM of simple ciliated columnar epithelium lining the bronchioles. The epithelium is composed of simple tall columnar cells with cilia projecting into the lumen. A few goblet cells are present among the columnar cells. Goblet cells are absent in the terminal bronchiole. (1000×)

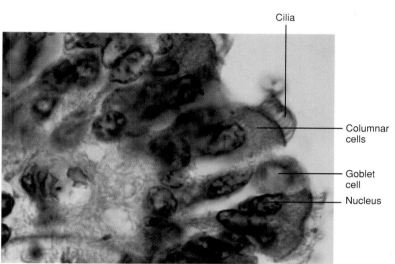

Cilia

Columnar cells

Goblet cell

Nucleus

Stratified Epithelium

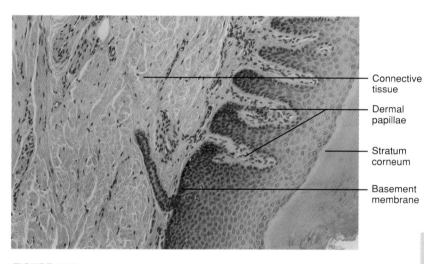

Connective tissue

Dermal papillae

Stratum corneum

Basement membrane

FIGURE 2.17
LM of stratified, keratinized squamous epithelium as seen in a thick skin. The epithelium is several layers thick and is constantly undergoing morphological changes. The basal stratum germinativum cells undergo mitotic division, forming new cells. As the old cells are pushed toward the periphery, they die and form the outermost covering, the stratum corneum. The basement membrane separates the epithelium from the underlying connective tissue. At intervals, the connective tissue projects into the epithelial layer, forming dermal papillae. (200×)

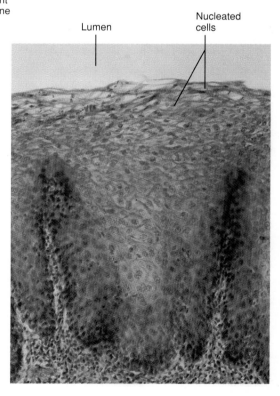

Lumen

Nucleated cells

FIGURE 2.18
LM of vaginal nonkeratinized stratified squamous epithelium. This epithelium is similar to esophageal epithelium. All cells of the epithelium are nucleated. The stratum corneum is composed of several layers of squamous-shaped cells that line the lumen of the vagina. (200×)

Bowman's
membrane

Stratified
epithelium

Fibroblast

Connective
tissue

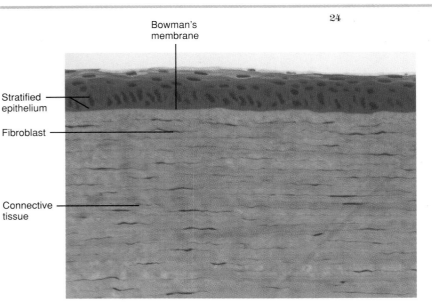

FIGURE 2.19
LM of the cornea of the eye, generally lined by
five to six layers of nonkeratinized stratified
squamous epithelium cells. The outermost layer
of the epithelium can easily be mistaken for a
simple squamous epithelium. However, lack of a
basement lamina (Bowman's membrane) identifies this layer as a part of a stratified epithelium.
Subepidermal connective tissue is infiltrated with
specialized fibroblast cells that lie between the
fibers. (400×)

Dermal
papillae

FIGURE 2.20
LM of upper region of esophagus, showing an internal
lining of nonkeratinized stratified squamous epithelium.
The basal cellular layers are composed of tall cells, the
intermediate layers of cells are polygonal in shape, and
the upper cell layers are flat and degenerating. The junction between the upper epithelial and lower connective
tissue is irregular, forming dermal papillae. (200×)

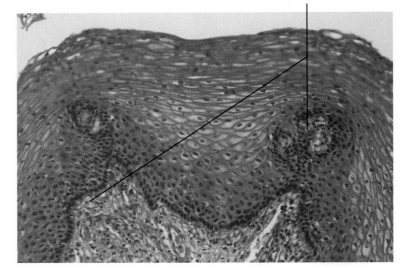

Cilia

Goblet
cells

Nucleus

Connective
tissue

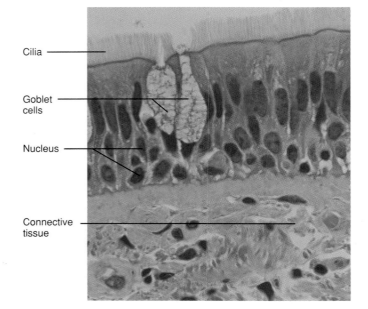

FIGURE 2.21
LM of pseudostratified ciliated columnar epithelium with interspersed goblet cells in the human trachea. This type of epithelium
is confined to the respiratory system; therefore, it is also called respiratory epithelium. It is pseudostratified because the cells are not
of equal height and present a stratified appearance even though
the basal portion of every cell rests on the basement lamina.
(1,000×)

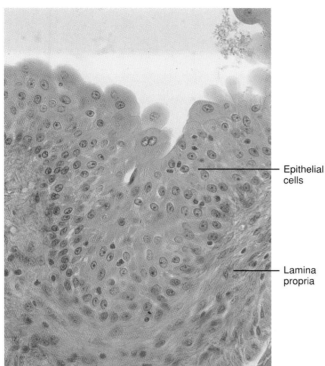

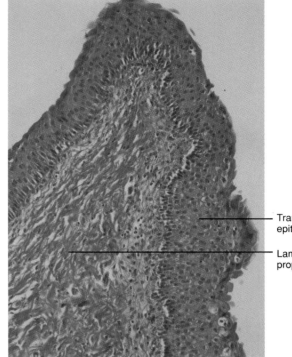

Epithelial
cells

Lamina
propria

Transitional
epithelium

Lamina
propria

FIGURE 2.22
LM of transitional epithelial lining of the urinary bladder.
Below the epithelium is the connective tissue region or
lamina propria. The epithelium flattens and changes
shape when the bladder is distended as a result of the
filling of the lumen. When the bladder is relaxed, the
cells appear to be thin and assume the shape of stratified squamous epithelium. The transitional epithelium
(from squamous epithelium to cuboidal) forms an
osmotic barrier that is impermeable to the diffusion of
water and salts. (400×)

FIGURE 2.23
LM of transitional epithelium which forms the inner
lining of the urinary bladder. As in the preceding
micrograph, the epithelial cells change shape from
low squamous to high cuboidal with distention and
contraction of the urinary bladder. In histological
preparation, the epithelium always appears to be
stratified squamous. In the micrograph, the lamina
propria shows distinct bands of collagen fibers.
(200×)

Epithelium

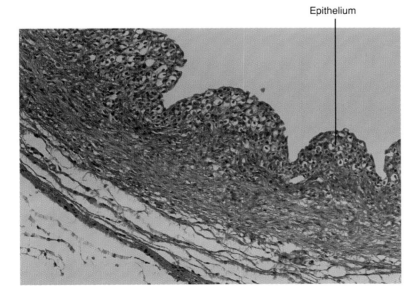

FIGURE 2.24
LM of transitional epithelium lining the ureters. The cellular morphology is similar to that of the lining of the urinary bladder. As in the case of the urinary bladder, the
epithelium serves as an impermeable barrier between
the luminal filtrate and the ureter body wall. The lamina
propria connective tissue has a higher concentration of
fibroblast cells than the deeper layer of connective
tissue. (200×)

Epithelium

Lamina propria

Stereocilia

FIGURE 2.25
LM of tall pseudostratified columnar epithelial cells in the lining of the epididymis. The principal cells bordering the luminal surface have been modified to form nonmotile microvilli or stereocilia. The surrounding lamina propria contains collagen and elastic fibers. (400×)

FIGURE 2.26
LM of the pseudostratified columnar epithelium lining of the ductus deferens, resting on a thin basement lamina. The cells bordering the lumen have stereocilia. The underlying lamina propria connective tissue exhibits dense collagen and elastic fibers. (1000×)

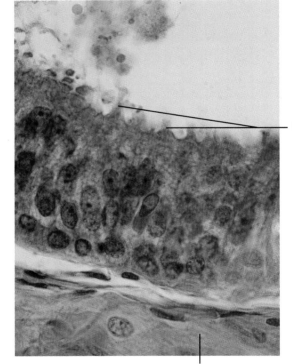

Stereocilia

Lamina propria

Cuboidal cells

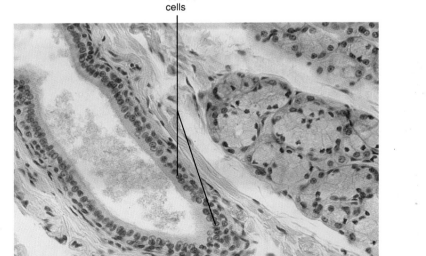

FIGURE 2.27
LM of stratified cuboidal epithelium in the ducts of a submaxillary gland. The epithelium is generally found in large exocrine glands and consists of two or three layers of low columnar or cuboidal cells. (400×)

Exocrine Glands

Glands are classified on the basis of their secretion and shape. If the ducts of the gland are not branched, the gland is called **simple.** If the ducts are branched, the gland is classified as a **compound** gland. Exceptions to this classification are **goblet cells,** which are unicellular, mucus-secreting exocrine glands associated with the epithelial lining.

1. **Simple tubular glands.** A gland that exhibits a single unbranched duct is designated as a simple tubular gland. Examples of such glands can be found in the crypts of Lieberkühn in the gastrointestinal tract.

2. **Simple coiled tubular glands.** A simple coiled tubular gland has a long duct. The only examples are the sweat glands, which have a terminal excretory portion lined with epithelial cells and a tubular section lined with stratified cuboidal cells.

3. **Simple branched tubular glands.** Each of these glands may form several tubular secretory units that secrete their products into a single tubular duct. Such glands are found in the lining of the stomach.

4. **Simple acinar glands.** These glands form clusters of secretory cells embedded in the epithelial lining. Such glands can be found in the penile urethra.

5. **Compound tubular glands.** This type of gland is also called a **compound tuboalveolar gland.** The gland is composed of several terminal, rounded or elongated secretory units called **alveoli.** The secretory cells of the alveoli are pyramidal or columnar in shape. Ducts from the alveoli merge to form a common large duct that opens into the lumen. Major salivary glands—parotid, submaxillary, and sublingual—are examples of compound tubular glands.

6. **Compound alveolar glands.** The terminal secretory units in these glands are also rounded or alveolar in shape, but the central lumen is large and is surrounded by low columnar or cuboidal secretory cells. The alveoli open into small ducts, which then

join to form larger ducts. Mammary glands are good examples of compound alveolar glands.

Endocrine glands release their secretory products, in the form of hormones, into the bloodstream, whereas the exocrine glands release their secretory products into a well-developed system of ducts which open on the body surface, or into the lumen of the gastrointestinal tract. Many hormone-secreting glands, such as pineal, pituitary, thyroid, parathyroid, and adrenal glands, and clusters of endocrine cells associated with other organs of the body, form the basis of the endocrine system. The endocrine system will be covered extensively in a subsequent chapter.

FIGURE 3.1
Light micrograph (LM) of goblet cells in the inner epithelium lining of the small intestine. Goblet cells are unicellular glands that secrete mucus and are found scattered in simple epithelial linings of the gastrointestinal and respiratory tracts. Each goblet-shaped cell contains a concentration of mucigen granules that form a thick mucus when mixed with water. Mucigen is a combination of neutral and acidic mucopolysaccharides (proteoglycans). The nucleus and organelles of the goblet cell are located in the stem part of the cell. (200×)

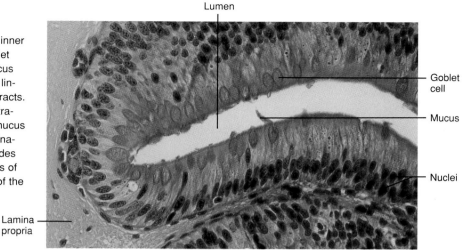

Lumen — Goblet cell — Mucus — Nuclei — Lamina propria

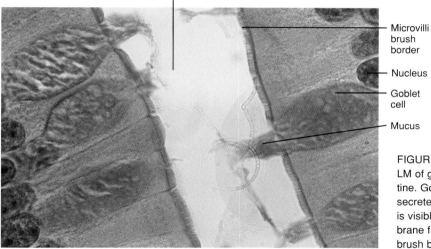

Lumen — Microvilli brush border — Nucleus — Goblet cell — Mucus

FIGURE 3.2
LM of goblet cells in the inner lining of the small intestine. Goblet cells are unicellular exocrine glands that secrete a thick viscous substance, called mucus, which is visible in this micrograph. The columnar cell membrane facing the lumen is modified to form a microvilli brush border. (400×)

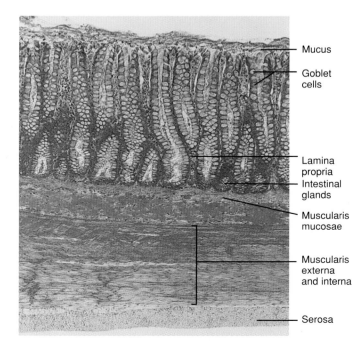

Mucus — Goblet cells — Lamina propria — Intestinal glands — Muscularis mucosae — Muscularis externa and interna — Serosa

FIGURE 3.3
LM of colonic (colon) crypts as seen in the mucosa. The linings of goblet cells on the walls of the crypts are good examples of simple tubular glands. The goblet cells secrete mucus, which protects the luminal surface of the tunica mucosa. Below the mucosa, the connective tissue of the tunica submucosa, the muscularis, and the tunica serosa can be identified. (100×)

Lumen of large intestine Goblet cells

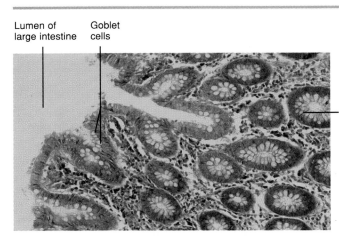

Gland cells

FIGURE 3.4
LM of a transverse section passing through simple tubular glands of colic (colon) mucosa. The goblet cells form a rosette of mucus-secreting cells. The cells are surrounded by the connective tissue of the lamina propria. The lumen of each crypt can be identified in the center of the goblet cell arrangement. (200×)

Goblet cell

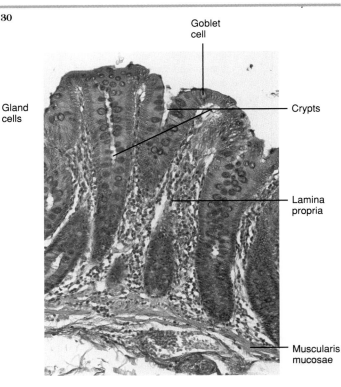

Crypts

Lamina propria

Muscularis mucosae

FIGURE 3.5
LM of simple tubular glands of colic (colon) mucosa, at a higher magnification (compare Figure 3.3). This micrograph shows the close relationship between the goblet cells and the crypts. Note the lamina propria of the tunica mucosa and the connective tissue of the tunica submucosa (200×).

Gastric epithelium

Gastric pits

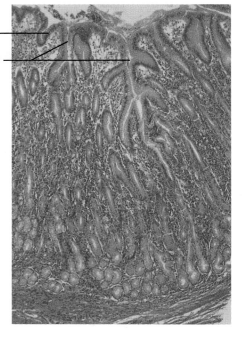

FIGURE 3.6
LM of simple branched tubular glands which are confined to the mucosa of the pyloric stomach. The branched tubular structures are lined by low columnar mucus, zymo-genic, parietal, and enteroendocrine cells. Gastric pit fove-olae can be seen in the micrograph. (100×)

Gastric epithelium

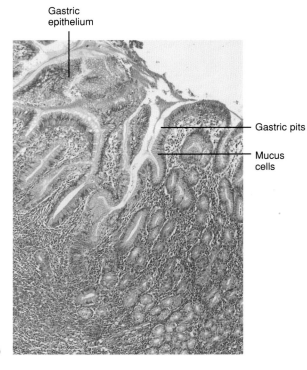

Gastric pits

Mucus cells

FIGURE 3.7
LM of a simple branched tubular gland of the pyloric mucosa at a higher magnification. In this micrograph, the pyloric pits and some of the cells lining the pits can be identified. Several branched excretory ducts are visible in this micrograph. (200×)

FIGURE 3.8

LM of a simple coiled tubular sweat gland of the skin. The gland consists of a single tube that is coiled at the base. In a cross section, the tubular structure is seen in various planes. The terminal secretory portions are identified in the micrograph. (400×)

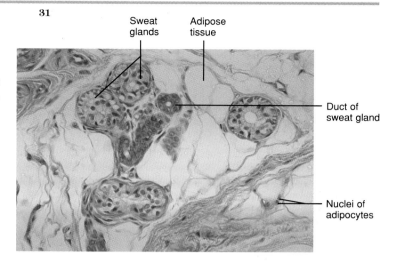

Sweat glands

Adipose tissue

Duct of sweat gland

Nuclei of adipocytes

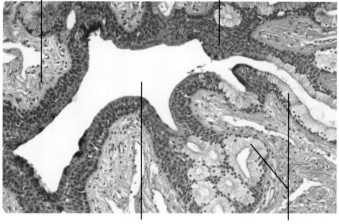

Corpora cavernosum urethrae

Pseudostratified columnar epithelium

Lumen

Paraurethral glands

FIGURE 3.9

LM of simple acinar glands in the epithelium of the male penile urethra. These glands form clusters of secretory cells that are embedded in the epithelial lining. (400×)

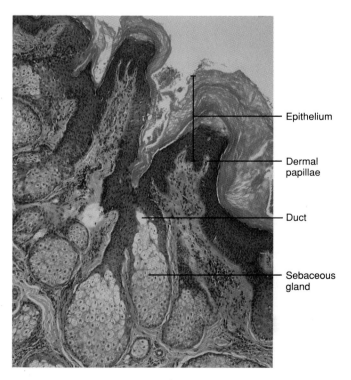

Epithelium

Dermal papillae

Duct

Sebaceous gland

FIGURE 3.10

LM of a sebaceous gland of the skin, classified as a simple branched alveolar gland. Each gland may consist of several secretory acini. Excretory products from the acini are emptied into a single excretory duct. This process of secretion is classified as holocrine, since entire cells are released as secretory vesicles. (100×)

Sebaceous
gland cells

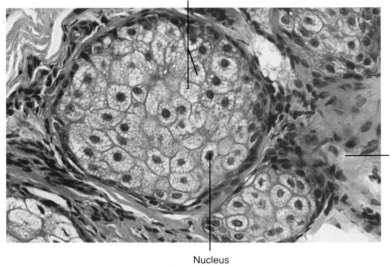

FIGURE 3.11
LM of a sebaceous gland of the skin magnified to show
the secretory cells of the acini. The entire cell ruptures,
releasing sebum as excretory product into a single excretory duct. (400×)

Connective
tissue

Nucleus

FIGURE 3.12
LM of a compound tubular gland, also
called a tuboalveolar gland, as seen in a
cross section of the submandibular gland.
The gland is composed of several terminal
rounded or secretory units called alveoli.
The secretory cells of the alveoli are pyramidal or columnar in shape. Ducts from the
alveoli merge to form a common large duct
that opens into the lumen of the oral cavity.
(400×)

Connective
tissue

Myoepithelial
cell

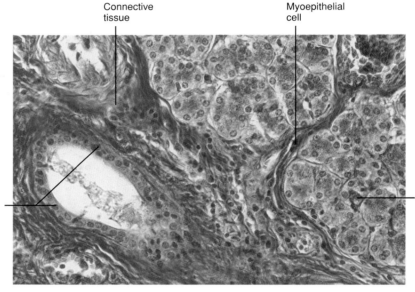

Salivary
duct

Secretory
granules

FIGURE 3.13
LM of a compound tubuloacinar gland as
seen in a cross section of the parotid salivary gland. Because the secretory units
are more rounded, the gland is called an
acinar or alveolar gland. The secretory
portion is classified as acinous. The elaborate and branched ducts demonstrate that
it is a compound gland. (400×)

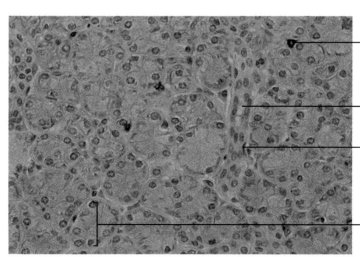

Small
duct

Interlobular
excretory duct

Myoepithelial
cell

Serous
acini

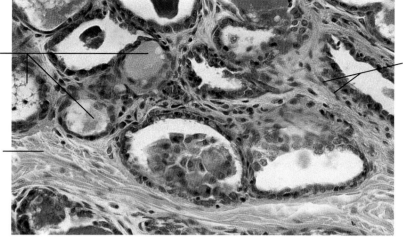

Branching alveoli with secretion

Interlobular connective tissue

Myoepithelial cells

FIGURE 3.14
LM of a compound alveolar gland as seen in a cross section of the mammary gland. The terminal secretory units in these glands are also rounded or alveolar in shape, but the central lumen is large and is surrounded by low columnar or cuboidal secretory cells. The alveoli open into small ducts, which then join to form larger ducts. (400×)

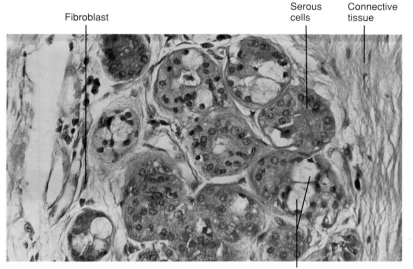

Fibroblast

Serous cells

Connective tissue

Mucous cells

FIGURE 3.15
LM of a compound tubuloacinar gland in the connective tissue of the trachea. The gland consists of a mixture of serous and mucus acini. The mucus acini cells are lightly stained and the cytoplasm has an almost colorless appearance. The nuclei are flat and are near the cell membrane. The serous acini cells stain much darker than the mucus acini cells. The cells are pyramidal or columnar in shape and are arranged around a small lumen of the alveolus. (400×)

FIGURE 3.16
LM of a mucus-secreting Brunner's gland of the small intestine (duodenum), classified as a compound branched tubular gland. The duct system is branched, and the secretory cells are composed of low columnar cells that secrete mucus. (400×)

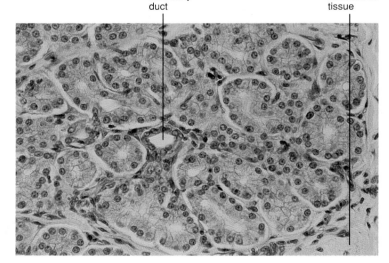

Secretory duct

Connective tissue

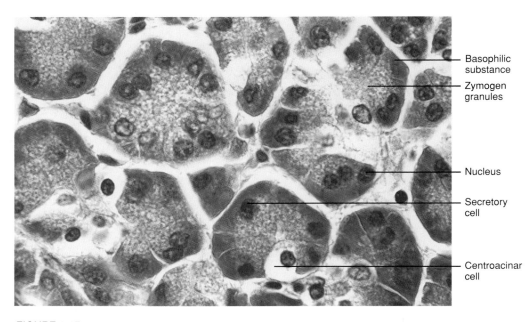

Basophilic substance

Zymogen granules

Nucleus

Secretory cell

Centroacinar cell

FIGURE 3.17
LM of a compound acinar gland in the pancreas. The secretory units drain into a well-organized, branched duct system. The micrograph shows several acini with zymogenic secretory cells bordering the acinus. The lumen of the acini are filled with zymogen granules. (1000×)

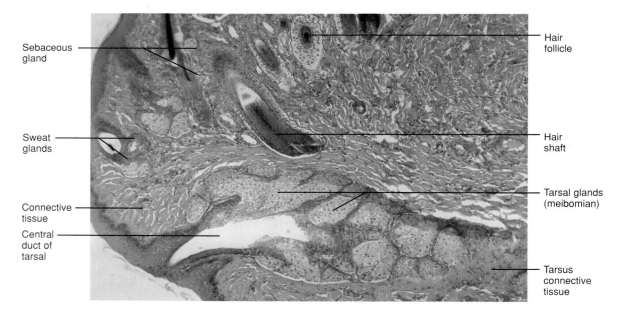

Sebaceous gland

Sweat glands

Connective tissue

Central duct of tarsal

Hair follicle

Hair shaft

Tarsal glands (meibomian)

Tarsus connective tissue

FIGURE 3.18
LM of tarsal (meibomian) glands in the tarsal plate of the eyelid. The tarsal glands are modified sebaceous glands with large numbers of acini. The acini secrete products into a long central duct that opens into the eyelid margin. (100×)

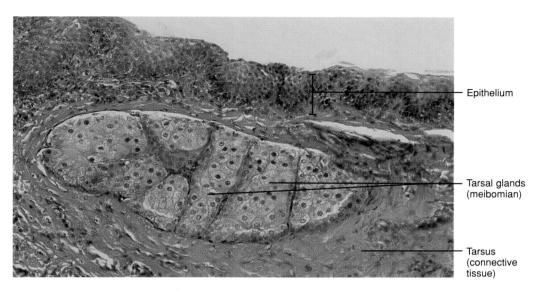

FIGURE 3.19
LM of tarsal (meibomian) glands of the eyelid at higher magnification. The glands are composed of several sebaceous acini. Secretions from the glands serve to lubricate the surface of the lid. (200×)

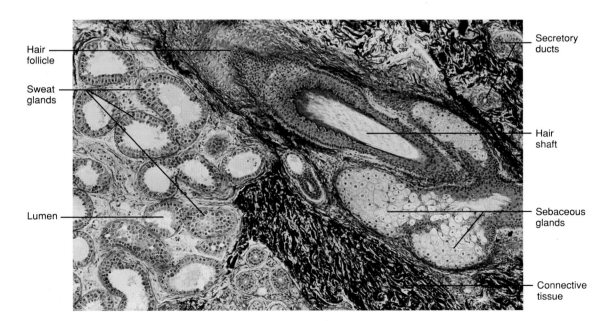

FIGURE 3.20
LM of modified sweat glands in the axilla. Although most sweat glands are classified as eccrine or merocrine (glands that do not lose considerable amounts of their cytoplasm during the secretory stage), some investigators classify axillary sweat glands as apocrine (gland cells that lose part of their cytoplasm during the secretory stage). The axillary glands are less coiled, are larger in size, and have lumina that are widely dilated in comparison with those of ordinary sweat glands. The cells are short columnar or high cuboidal, and the cell nuclei are prominent and stain deeply. The glands and ducts are separated by dense collagen connective tissue. The glands are surrounded by myoepithelial cells. Contraction of these cells facilitates secretion. In the micrograph, sebaceous glands and a hair shaft are located close to the axillary glands. The axillary glands are found in the axilla and the pubic area of the body. They become active at puberty. The secretory products from these glands are rich in organic matter. When these products interact with bacteria, they produce an offensive odor. Emotional stress, exercise, and adrenergic nerve stimulation can enhance the secretory activity of the axillary glands. (100×)

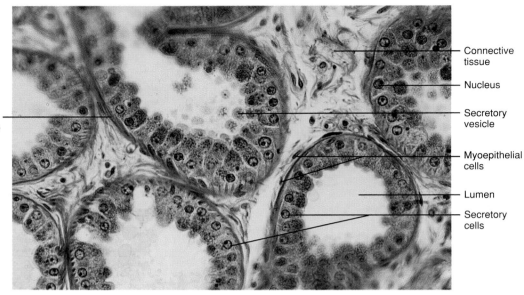

Connective tissue

Nucleus

Secretory vesicle

Myoepithelial cells

Lumen

Secretory cells

Basement membrane

FIGURE 3.21

LM of axillary sweat glands magnified to show the secretory cells and their mode of secretion. The cells are budding secretory vesicles that enclose part of the cellular cytoplasm. The budding of the apical cytoplasm gives a false indication that the glands are apocrine; however, recent studies indicate that this appearance may be the result of a fixation artifact. As in the preceding micrograph, the glands are surrounded by a well-defined basement membrane, collagen fibers, and myoepithelial cells. In humans, the biological importance of these glands is not clear. The glands are similar to the odiferous glands of mammals. (400×)

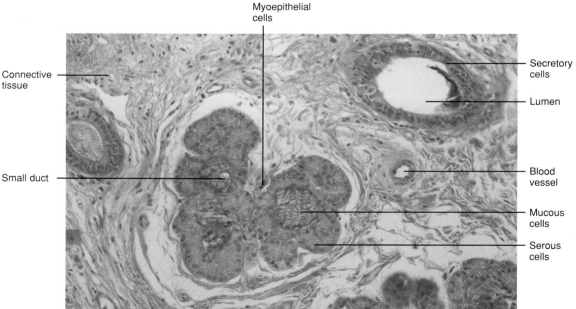

Myoepithelial cells

Secretory cells

Lumen

Connective tissue

Small duct

Blood vessel

Mucous cells

Serous cells

FIGURE 3.22

LM of mucus-secreting glands located in the lamina propria of the vocal folds. These glands are seromucous or mixed glands. The serous cells stain darker than the mucous cells. In the micrograph, the glands are surrounded by collagen fibers. Two secretory ducts, myoepithelial cells, and a blood vessel are also visible in this micrograph. (200×)

Connective Tissue

In a developing embryo, the majority of the connective tissue is derived from mesenchyme, a derivative of the mesodermal germinal layer. The connective tissue is a combination of a diverse group of tissues that are responsible for performing several functions. Even though these tissues are structurally and functionally diverse, they share common qualities and for this reason are considered collectively. Functionally, connective tissues provide support, transport, defense, repair, storage, packing material, and insulation. Some of the functions may be general whereas other functions may be delegated to specialized connective tissues such as blood, cartilage, and bone.

On the basis of tissue structure and function, the connective tissues can be classified in several ways. The following general classification is widely accepted.

I. **Fetal Connective Tissues**
   A. **Mucus:** e.g., umbilical cord
   B. **Mesenchymal:** e.g., developing embryo and fetus

II. **Adult Connective Tissues**
   A. **Loose or areolar:** packing tissue associated with most organs

B. **Reticular:** tissue found in bone and lymph nodes
   C. **Adipose:** storage tissue in subcutaneous and omentum areas
   D. **Dense irregular:** found in dermis, periosteum, perichondrium, and capsules of organs
   E. **Dense regular:**
      1. **Collagenous:** e.g., tendons, aponeurosis, cornea, and ligaments
      2. **Elastic:** e.g., ligamentum nuchae, ligamentum flavum, and suspensory ligament of penis

III. **Specialized Connective Tissues**
   A. **Supporting tissue—cartilage:**
      1. **Hyaline:** e.g., in trachea and costal cartilage
      2. **Elastic:** e.g., in epiglottis and external ear
      3. **Fibrous:** e.g., symphysis pubis, intervertebral disk
   B. **Supporting Tissue—bone:**
      1. **Cancellous,** or spongy bone
      2. **Compact,** or dense bone

IV. **Transport Tissues:** e.g., blood, cardiovascular system and hemopoietic tissue

Connective Tissue Cells

A diverse group of cells is found in the connective tissue. Cells of the loose or areolar tissue are organized into six basic categories.

Fibroblasts are fusiform cells which are abundant in connective tissue. The cells are active in the synthesis of collagen, reticular, and elastic fibers. The cells secrete most of the ground substance, which is chemically a mixture of mucopolysaccharides. Included are hyaluronic acid, chondroitin sulfate, and undifferentiated substances.

Macrophages (histiocytes) are derivatives of monocytes. These are the main phagocytosing cells of the connective tissue and are responsible for the ingestion of foreign particulate matter. Macrophages have a close relationship with the lymphocytes and are important in the immunological process.

Plasma cells are common in connective tissue, where their function is to bring about humoral immunity by secreting antibodies. Plasma cells are derivatives of "B" lymphocytes.

Mast cells are not as common in connective tissue as the previously mentioned cells. Generally, mast cells are found in the vicinity of blood vessels. Mast cells have metachromatic granules which store histamine, a vasodilator, and heparin, an anticoagulant.

Fat (adipose) cells may be present individually or in small clusters in the connective tissue. Adipose cells store lipids and provide insulation and protection.

Leukocytes (white blood cells) migrate from the blood vascular system into the connective tissue by diapedesis. Their function is primarily to defend against bacteria and other foreign organisms. Neutrophils are effective phagocytes; they increase in number at sites of bacterial infection. The eosinophils increase in number in the connective tissue when there is a parasitic infection or allergic reaction. Eosinophils are also responsible for ingestion of antigen antibody complexes formed during allergic reactions.

Connective Tissue Fibers

Based on their morphological differences, connective tissue fibers can be divided into three groups: collagen, elastic, and reticular fibers.

Collagen fibers are the most common fibers of the connective tissue and are found in almost all types of connective tissue. Collagen fibers give great tensile strength to organs and parts of the body where skeletal movement occurs.

Elastic fibers are less common than collagen fibers. They are thin, branched, and relatively short, and they have lower tensile strength. Elastic fibers exhibit elasticity: they can stretch and return to their initial size without any physical distortion. These fibers are quite common in the lungs, large blood vessels, urinary bladder, skin, and elastic cartilage.

Reticular fibers are the least common of all the connective tissue fibers. They are found in the lymph nodes, spleen, liver, and hemopoietic tissue. The reticulate fibers can be delineated using silver stains.

Ground Substance of Connective Tissue

The ground substance forms a permeable amorphous matrix in which fibers and cells are embedded. Chemically, the matrix is a mixture of proteoglycans, which creates a gel-like consistency, and glycosaminoglycans (carbohydrate polymers). It also contains hyaluronic acid, chondroitin-6-sulfate, chondroitin-4-sulfate, dermatan sulfate, and heparin sulfate. Glycoprotein and fibronectin may also be present in the connective tissue proper.

FIGURE 4.1
Light micrograph (LM) of mucous connective tissue as seen in the fetal umbilical cord (Wharton's jelly). The matrix of mucous tissue contains loosely arranged collagen fibers with large numbers of fibroblast cells. The fibroblast cells have centrally placed nuclei and fusiform-shaped cytoplasm. (400×)

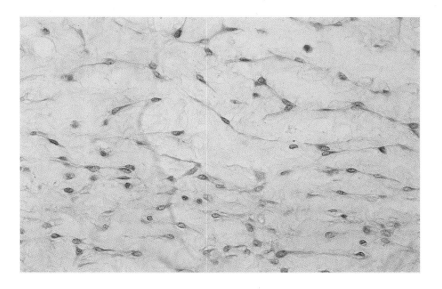

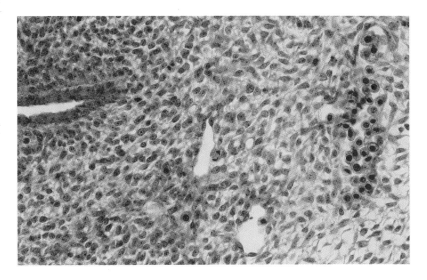

FIGURE 4.2
LM of mesenchymal connective tissue in a human fetus. The tissue is still at an immature state and is partially cellular. The mesenchymal cells undergo differentiation before developing into any given organ. The mesenchymal cells are fusiform to stellate-shaped. The cytoplasm gives a definite outline to the cells, and the ground substance fills the spaces between cells. The hollow areas are formed by the extraction of ground substance. A blood vessel can be seen at upper left. (400×)

FIGURE 4.3
LM of reticular connective tissue in the medulla of a lymph node. The fibers are thin, branched, and fairly long. In the lymph node, these fibers form a filtering network. Lymphoid cells, lymphocytes, reticulocytes, and macrophages are closely associated with these fibers. Also visible in the micrograph are trabeculae and trabecular blood vessels. (200×)

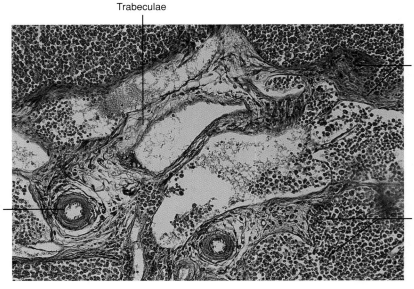

Trabeculae

Reticular fibers

Trabecular blood vessel

Lymphocytes

Elastic fibers   Fibroblast

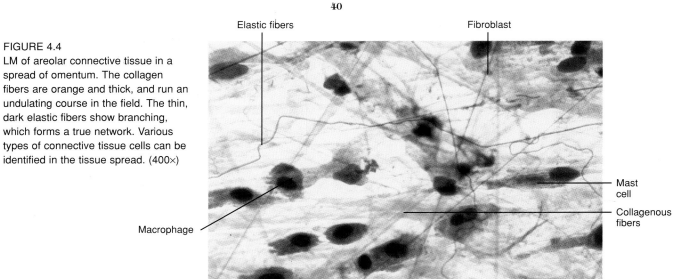

Mast cell

Collagenous fibers

Macrophage

FIGURE 4.4
LM of areolar connective tissue in a spread of omentum. The collagen fibers are orange and thick, and run an undulating course in the field. The thin, dark elastic fibers show branching, which forms a true network. Various types of connective tissue cells can be identified in the tissue spread. (400×)

FIGURE 4.5
LM of adipose connective tissue, showing adipocytes, or fat cells without any cytoplasmic details. The cytoplasm appears to be concentrated in the peripheral area of the cell. The nucleus is also pressed toward one side of the cell. In the living state, the adipocytes are filled with lipid molecules. (400×)

Fibroblast   Venule

Nucleus

Capillary

Adipose cells

Nucleus   Adipocytes and lobules

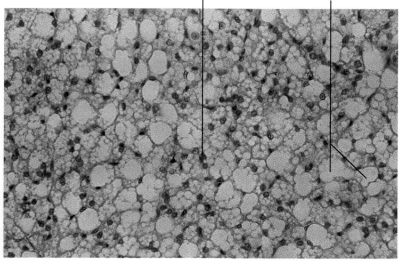

FIGURE 4.6
LM of brown adipose tissue, generally found in newborn and hibernating animals. The adipocytes have round nuclei, and the cytoplasm is filled with small fat droplets. For this reason, this type of fat is also known as multilocular adipose tissue. (200×)

FIGURE 4.7
LM of dense, regular connective tissue, such as in tendons, contains a predominance of collagen fibers in parallel rows. Between the fibers are small blood vessels and collagen-producing fibroblast cells. In this micrograph, the nuclei of the fibroblasts are visible. The undulating arrangement of the collagen fibers is characteristic of tendons. (200×)

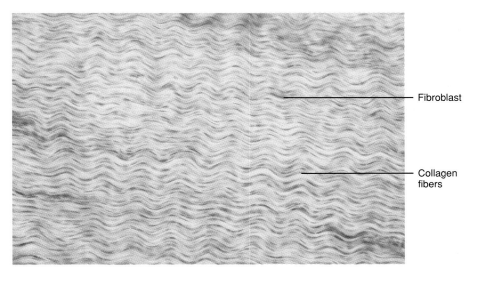

Fibroblast

Collagen fibers

Desquamating layers of epidermis

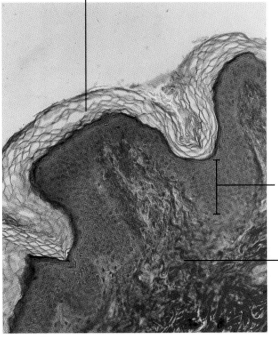

FIGURE 4.8
LM of dense, irregular connective tissue in the dermis of the skin. The tissue contains compact masses of collagen fibers that run in irregular directions. (200×)

Epidermis

Collagen fibers

FIGURE 4.9
LM of dense, regular connective tissue in a longitudinal section of ligamentum nuchae. There are a few fibroblast nuclei between the fibers. The elastic fibers are branched, broad, and nearly parallel. Very few fibroblast cells are present between fibers. (200×)

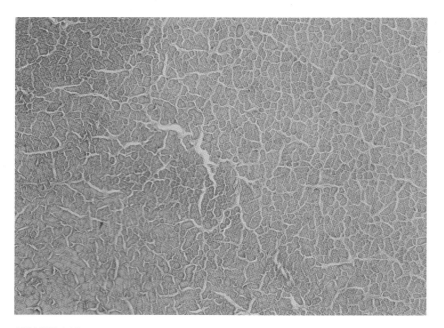

FIGURE 4.10
LM of dense, regular elastic connective tissue in a cross section of ligamentum nuchae. Between the elastic fibers, which are broad, sparse, and delicate, are homogeneously distributed collagen fibers. In the fibroblasts, a few nuclei are visible between the elastic and collagen fibers. (200×)

FIGURE 4.11
LM of elastin connective tissue as seen in the body wall of the aorta. Elastin, a protein, is generally found in all connective tissue in varying amounts. In this micrograph, the fibers are wavy, branched, and darkly stained by van Gieson's stain. (200×)

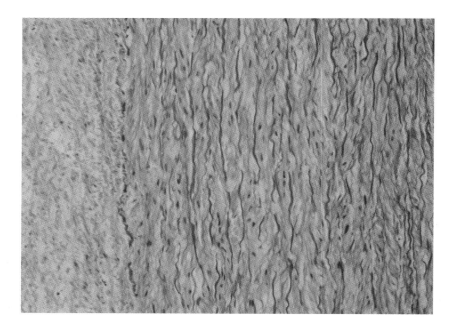

Specialized Connective Tissue: Cartilage

Skeletal tissue, which includes **cartilage** and **bone,** is specialized connective tissue composed of fibers, cells, and ground substance, the latter two of which form the intercellular matrix. Chemically, the ground substance in cartilage is composed of chondromucoids that have a high concentration of sulfated proteoglycans and hyaluronic acid. The combination of the two forms chondronectin and aggregates of proteoglycans. In comparison, collagenous fibers in bones are impregnated with an amorphous ground substance composed predominantly of calcium hydroxide crystals.

In early embryonic development, cartilage forms the framework of the skeleton. In the late stages of development, most of the cartilage is replaced by bone. Exceptions to this are the articulating surfaces of bone, parts of the respiratory system, parts of the ear, and junctional areas between the ribs and the sternum.

The **cartilage matrix** is composed of elastic and collagen fibers that increase the flexibility and strength of the tissue so that it can adapt to stress and the mechanical requirements of different regions of the body. Further, the extracellular matrix, with its connective tissue and ground substance, modifies the tissue to form a flexible semirigid gel. This increases the pliability of the tissue, which can then conform to any physical changes that may take place in the body.

Cartilage is different from other connective tissue in that it does not have blood vessels, nerves, or lymph vessels. It is entirely dependent on the blood vessels in the surrounding tissue for its nourishment. The nutrients are transported by diffusion to the developing cartilage cells.

Cartilage, unlike other forms of connective tissue, has secretory cells called **chondrocytes,** which are embedded in the intercellular matrix. Surrounding the cells are empty spaces called **lacunae.** However, the space around the cells in reality is a fixation artifact caused by the shrinking of the matrix at the time of tissue processing. The cells found in the peripheral region of the cartilage, below the perichondrium, are called **chondroblasts.** The chondroblasts/chondrocytes create an elaborate extracellular matrix. Depending on the type of cartilage, the matrix may have type I or type II collagen fibrils with 64-nm periodicity, or may have a predominance of elastic fibers. On the basis of abundance and type of fibers present in the matrix, cartilage can be classified into three types: **hyaline cartilage, elastic cartilage,** and **fibrous cartilage.** The three types of cartilage are distributed in different parts of the body according to function.

Hyaline Cartilage

Hyaline cartilage is one of the most common types of cartilage and is found as costal cartilage between the ribs and the sternum, articular cartilage between bone joints, and tracheal rings of the trachea and bronchi. It is also the cartilage of the nose and larynx, and the cartilage that proliferates in elongation of areas of the long bones. Surrounding the hyaline cartilage is a dense mass of connective tissue that forms the covering or the perichondrium.

Chondrocytes in the lacunae may be uninucleate or multinucleate. As the cartilage grows, cells deep in the cartilage become more rounded, whereas cells below the perichondrium are elliptical in shape with a long axis lying parallel to the surface. At times, the chondrocytes can be found in small clusters inside lacunae, forming isogenous groups of cells that, in fact, are chondrocytes undergoing mitotic division.

In routine histological preparations, or in a fresh specimen, the matrix appears to be relatively homogeneous with collagen fibers assimilated into the ground substance. The ground substance is mainly proteoglycans (chondroitin-4 and chondroitin-6 sulfate, keratin sulfate, and a small amount of hyaluronic acid, giving a basophil property to the matrix) consisting of proteins and complex carbohydrates (glycosaminoglycans). The peripheral layer of the perichondrium is highly vascularized. It has dense collagen and elastic fibers and is infiltrated with fibroblasts. The underlying chondroblast cells secrete ground substance, and later, as they are trapped in their own matrix, these cells are called chondrocytes.

Elastic Cartilage

Elastic cartilage, which is a type of modified hyaline cartilage, has a predominance of branched elastic fibers that almost obscure the matrix. Collagen fibers, which are present in hyaline cartilage, are also present in elastic cartilage, but these fibers are completely masked by elastic fibers. Deep inside the cartilage, the thick, closely packed masses of elastic fibers obscure the ground substance. Just below the perichondrium, the elastic fibers are loosely arranged and form a network of fibers that blends with the overlying perichondrium. Owing to the nature of elastic fibers, elastic cartilage is more flexible than hyaline and fibrous cartilage. The elastic cartilage can be seen in such places as the auditory tube, the external ear, the epiglottis, and the smaller cartilages associated with the larynx.

Fibrous Cartilage

Unlike hyaline and elastic cartilage, **fibrous cartilage,** or **fibrocartilage,** is not a derivative of mesenchyme tissue. It is a product of dense connective tissue that has differentiated into cartilage as a result of stress and the weight-bearing demands of the body. Fibrous cartilage is found in the intervertebral disks, symphysis pubis of the pelvic girdle, menisci, and ligaments associated with joints and junctional areas between bones, tendons, and ligaments.

Fibrous cartilage consists of typical lacunae with cartilage cells or chondrocytes. It can be seen with small amounts of ground substance surrounding the cells. The cells may lie singly, in pairs, or in short rows of cells surrounded by collagenous fibers that exhibit a 64-nm banding pattern. In fibrous cartilage, the perichondrium is absent. Depending on the location of the fibrous cartilage, the fibers may blend into the surrounding hyaline cartilage, bone, or dense fibrous tissue.

Chondroid Tissue

In lower vertebrates during embryonic development, the hyaline cartilage may go through a transitory stage where the adjacent vessicular cells are surrounded by a thin capsule and collagenous fibers in its interstitium. This immature cartilage may remain and provide support for body parts.

Hyaline Cartilage

FIGURE 5.1
Light micrograph (LM) of fetal hyaline cartilage in a developing bone. Cartilage cells,
chondroblasts, and chondrocytes have not
clearly differentiated at this stage; undifferentiated cells can be seen throughout the
homogeneous matrix. Lacunae are small
or, in many cases, absent.The perichondrium is sparse around the chondrocytes;
and, at this stage, the cartilage zones have
not fully been established. (200×)

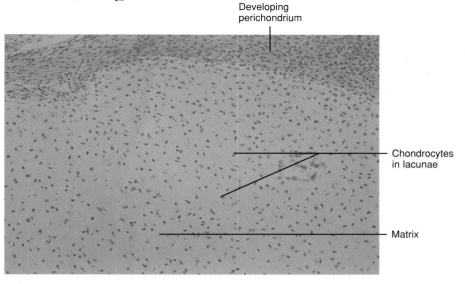

Developing
perichondrium

Chondrocytes
in lacunae

Matrix

FIGURE 5.2
LM of hyaline cartilage and the surrounding
tissue in a cross section of the trachea. The
perichondrium forms a distinct peripheral
covering of the cartilage. In the interior or
central region of the cartilage, lacunae with
chondrocytes can be seen throughout the
homogenous matrix. The matrix between
the lacunae—the interterritorial matrix—can
be identified. The matrix immediately surrounding the chondrocyte cells is called territorial matrix. Other tracheal tissues such
as surrounding connective tissue, tracheal
glands, glandular ducts, epithelium with
pseudostratified ciliated columnar cells,
lumen of the trachea, and blood vessels in
the connective tissue, can be clearly identified. (45×)

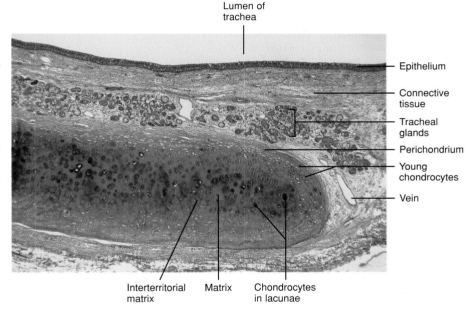

Lumen of
trachea

Epithelium

Connective
tissue

Tracheal
glands

Perichondrium

Young
chondrocytes

Vein

Interterritorial
matrix

Matrix

Chondrocytes
in lacunae

FIGURE 5.3
LM, at higher magnification, of the tracheal
hyaline cartilage plate with lacunae
throughout the homogeneous matrix. The
chondrocytes can be seen as single cells in
the lacunae or as isogenous groups inside
the lacunae. Perichondrium and blood vessels surround the cartilaginous matrix. In
among the lacunae is the interterritorial
matrix. (100×)

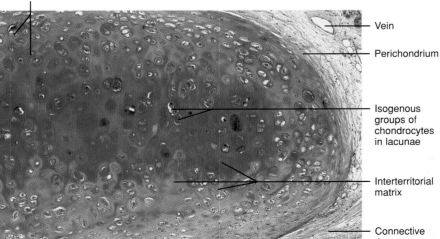

Young
chondrocytes

Vein

Perichondrium

Isogenous
groups of
chondrocytes
in lacunae

Interterritorial
matrix

Connective
tissue

Chondrocytes

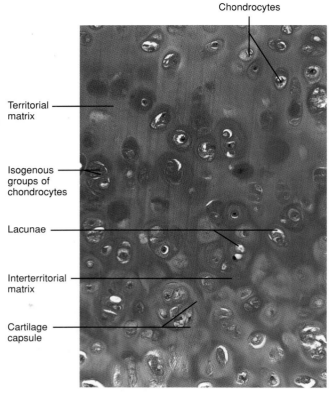

Territorial matrix

Isogenous groups of chondrocytes

Lacunae

Interterritorial matrix

Cartilage capsule

FIGURE 5.4

LM, at still higher magnification, of hyaline cartilage from the trachea, illustrating the chondrogenic layer where chondrocytes have differentiated from mesenchyme cells. Further proliferation of cells takes place by mitotic division. In this micrograph, isogenous groups of chondrocytes and single chondrocytes can be seen in the lacunae. Between the lacunae, collagen fibers have not fully been impregnated with ground substance. (200×)

Elastic Cartilage

Fibroblasts   Perichondrium   Small and large chondrocytes

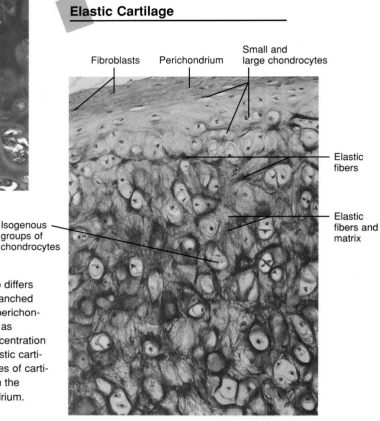

Elastic fibers

Elastic fibers and matrix

Isogenous groups of chondrocytes

FIGURE 5.5

LM of elastic cartilage taken from the epiglottis. This cartilage differs from other cartilages principally by the presence of dense, branched elastic fibers in its matrix. The elastic fibers extend from the perichondrium as small clusters of fibers, but proliferate in the interior as branching and anastomosing fibers of various sizes. The concentration and branching of fibers in the matrix vary among different elastic cartilages depending on the function of the organ. As in other types of cartilage, the interior of elastic cartilage has large chondrocytes in the lacunae and much smaller chondrocytes below the perichondrium. (200×)

Lacunae   Chondrocytes

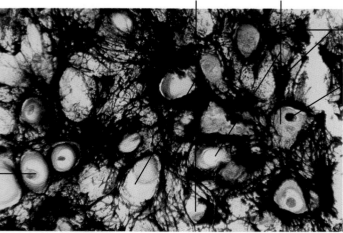

Elastic fibers

Nucleus

Chondrocyte with acentric nucleus

FIGURE 5.6

LM of elastic cartilage at a higher magnification. Dense clusters of elastic fibers weave around the lacunae. The elastic fibers are thick and compact; in some places they seem to obscure the matrix. Large chondrocytes can be seen in the lacunae. (400×)

Fibrous Cartilage

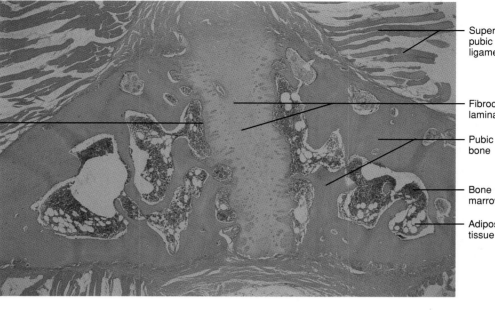

Superior pubic ligament

Fibrocartilaginous lamina

Pubic bone

Bone marrow

Adipose tissue

Hyaline cartilage

FIGURE 5.7
LM of fibrocartilage as seen in the symphysis pubis of the pelvis. The cartilage is composed of dense bundles of collagenous fibers. Lacunae are spread out in the matrix, and the cartilage lacks its own perichondrium. The peripheral regions of the cartilage blend into the surrounding ossified tissue. (40×)

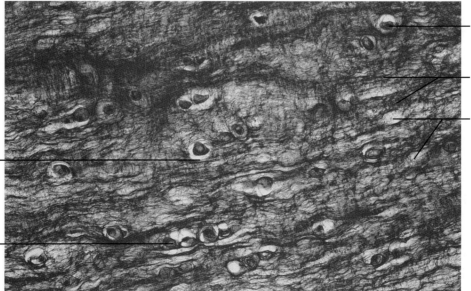

Nucleus of chondrocyte

Collagenous fibers

Matrix

Lacuna

Rows of chondrocytes

FIGURE 5.8
LM of fibrocartilage in the intervertebral disk. The cartilage shares similarities with dense fibrous connective tissue and fibrous cartilage. The cartilage tissue consists of dense collagenous fibers, chondrocytes, and a hyaline cartilage matrix. (100×)

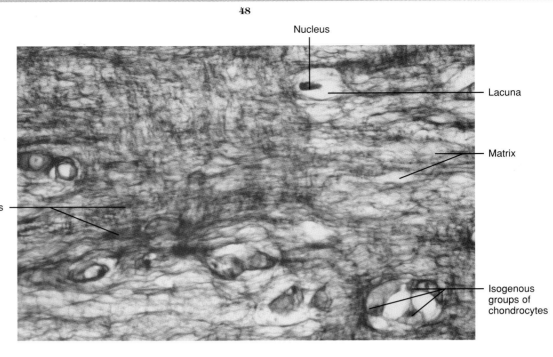

Nucleus

Lacuna

Matrix

Collagenous
fibers

Isogenous
groups of
chondrocytes

FIGURE 5.9
LM of fibrocartilage from an intervertebral disk at a higher magnification. Unicellular or bicellular
groups of chondrocytes can be seen in the matrix. Tufts of collagenous fibers surround the
lacunae. (200×)

Chondroid Tissue

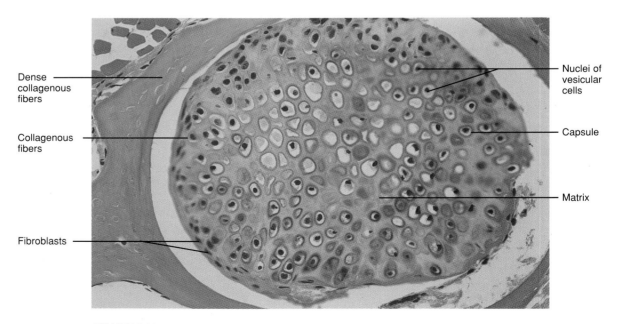

Dense
collagenous
fibers

Collagenous
fibers

Fibroblasts

Nuclei of
vesicular
cells

Capsule

Matrix

FIGURE 5.10
LM of a chondroid, sometimes described as a pseudocartilage—a primitive, early vertebrate
cartilage tissue found in cyclostomes. It consists of closely packed vesicular cells surrounded
by a capsule rich in collagenous fibers but poor in mucopolysaccharides. The tissue found in
notochords of vertebrates has a structure similar to that of chondroid tissue of the cyclostomes.
(200×)

Specialized Connective Tissue: Bone

Bone, as in the case of cartilage, is a modified connective tissue. However, the ground substance secreted by the bone cells is mineralized, resulting in a dense, hard, nonpliable, weight-bearing, and high-compression-strength substance. Bone is a complex living tissue that is highly vascularized and is constantly being deposited and reabsorbed throughout life. Hormones play important roles in the processes of bone deposition and bone reabsorption.

Bone matrix possesses a predominance of collagen fibers that are infiltrated with a heavily mineralized amorphous substance. Embedded in the bony matrix are bone cells or **osteocytes.** The osteocytes in bone (Gr. *osteon*), like the chondrocytes in cartilage, are surrounded by spaces or lacunae. Bone tissue also possesses a covering of dense connective tissue called periosteum. Bone cells, osteoblasts, and osteocytes are direct derivatives of mesenchyme, as is cartilage. However, the difference is that osteoblasts and osteocytes differentiate in the vicinity of blood capillaries, whereas chondroblasts

and chondrocytes differentiate in the subperichondrial region.

Another unique feature of bone is that its cells are nourished by blood capillaries that lie approximately 0.2 nm or less from the bone cells. The blood capillaries are incorporated into the matrix at the time of bone development. In addition to the blood capillaries, the bone matrix is traversed by fine channels, known as canaliculi, which are filled with nutritious fluid from the blood capillaries. The canaliculi connect the osteocytes, which are in the lacunae, to the source of nutrients, the blood capillary, thus providing a constant supply of oxygen and nutrition to all the osteocytes in bone tissue.

Osteogenesis (bone development) is brought about by two mechanisms: **intramembranous ossification** and **endochondral ossification.** Intramembranous ossification is confined to the formation of flat bones such as the cranial bones, mandible, facial bones, and the clavicles. The flat bones develop directly from vascularized mesenchyme in the embryonic stage. Endochondral ossification is much more complex. Initially, a cartilaginous framework, or cartilage model, is created. The cartilage model is gradually replaced by bony tissue in fetal life. Most of the axial (excluding skull) and appendicular bones are formed by endochondral ossification.

Cells associated with the bone remodeling of calcified bones are called osteoclasts. These cells are multinucleated and are formed by the fusion of numerous osteoprogenitor cells. The osteoclasts are generally present in bone depressions called Howship's lacunae. The periosteum that covers the external surface of the bone is heavily anchored to the bone by bundles of dense collagenous connective tissue fibers called Sharpey's fibers. The connective tissue is infiltrated by fibroblasts and osteoprogenitor cells.

Morphologically, bone can be classified into two types: **compact** or **dense bone,** and **cancellous** or **spongy bone.** In compact bone, the bone may be deposited either in layers parallel to one another or in a concentric manner around a blood vessel, thus presenting a lamellar arrangement of bone. This concentric lamellar deposition of bone around a blood vessel forms a microscopic unit comprised of a blood vessel, a lymph vessel, and possibly a nerve. This unit of bone is known as a Haversian system. Dense bone contains longitudinally oriented channels; these channels are remnants of blood vessels and are called Haversian canals. The Haversian canals communicate with each other by means of transverse and oblique channels. Still other channels, called Volkmann's canals, enter bone from the periosteal and endosteal surfaces and run perpendicular to the long axis of the bone. The Volkmann's canals in turn unite with the Haversian canals. In summary, the dense bone contains an extensive array of blood vessels and communicating canals that supply the nutritional needs of the bone tissue.

Cancellous (or spongy) bone is in many ways similar to dense (or compact) bone; however, the ossified trabeculae and spicules are thin and are surrounded by bone marrow and blood vessels. Osteons are mostly absent in spongy bone. The spaces between the spicules and trabeculae are filled with hemopoietic tissue.

In a typical long bone, the shaft, or diaphysis, is composed predominantly of compact bone that surrounds the medullary or bone marrow cavity. The terminal ends (epiphysis) of the long bones are essentially cancellous bone covered with a thin shell of compact bone. The marrow cavity of the long bone (diaphysis) is continuous with the marrow cavity of the cancellous bone.

Bone

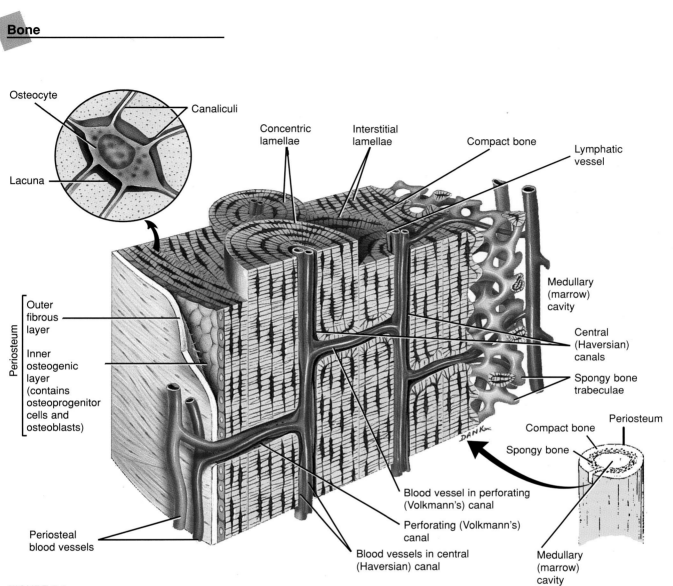

FIGURE 6.1
Diagram illustrating complex morphology of osteons in compact bone tissue. Supporting diagrams further illustrate such structures as the lacuna, spongy bone trabeculae, and a section of trabecula.

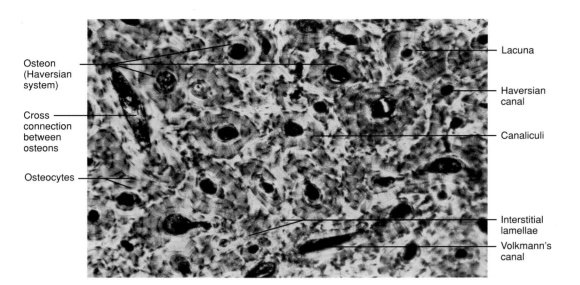

Osteon (Haversian system)

Cross connection between osteons

Osteocytes

Lacuna

Haversian canal

Canaliculi

Interstitial lamellae

Volkmann's canal

FIGURE 6.2
Light micrograph (LM) of ground bone from the diaphysis of a long bone. Various lamellar arrangements of a mineralized bone matrix are displayed. In the lamellae, osteocytes are confined to the almond-shaped lacunae. Radiating from the lacunae in all directions are fine channels called canaliculi, which anastomose and connect other lacunae and their osteocytes. Some canaliculi open into the Haversian canals and the bone marrow cavity. Haversian systems or osteons make up the structural units of bone. In a given osteon there are several concentric lamellae surrounding a central Haversian canal. In between the osteons are irregular interstitial lamellae. (100×)

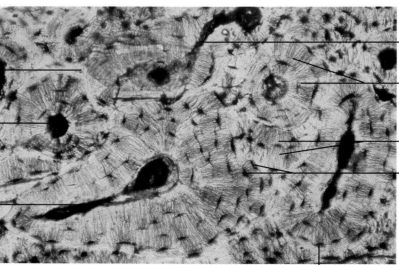

Matrix

Haversian canal

Volkmann's canal

Cross connection between Haversian canal

Canaliculi

Canaliculi of interstitial lamellae

Lacunae

Osteocytes in lacunae

FIGURE 6.3
LM of ground bone which has been made paper thin by grinding with abrasives. The canaliculi and almond-shaped lacunae with osteocytes are clearly demonstrated. The lacunae are arranged parallel to the osseous lamellae with canaliculi entering at right angles. The Haversian canals are connected with cross connections, the Volkmann's canals. (200×)

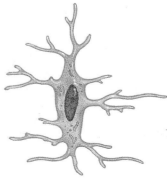

Osteocyte

Canaliculi

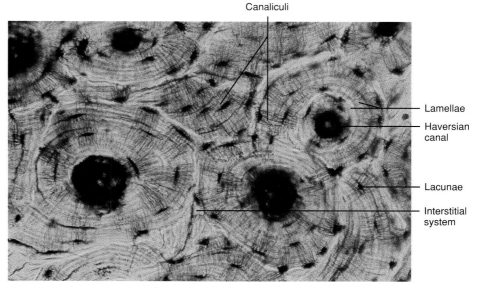

Lamellae

Haversian canal

Lacunae

Interstitial system

Canaliculi   Calcified matrix

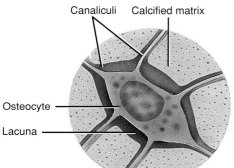

Osteocyte

Lacuna

FIGURE 6.4

LM of a Haversian canal in an osteon. In this micrograph, canaliculi radiate in all directions from the lacunae and connect with adjacent osteocytes and lacunae. The canaliculi form channels for the diffusion of metabolites and circulation of tissue fluids between the vessels of the Haversian canals and the lacunae. (200×)

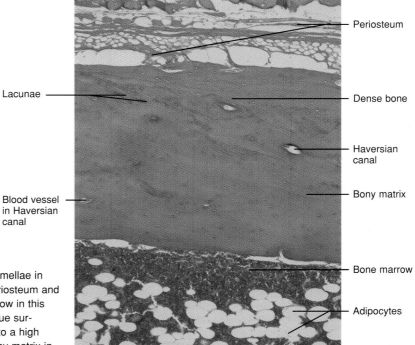

Periosteum

Lacunae

Dense bone

Haversian canal

Blood vessel in Haversian canal

Bony matrix

Bone marrow

Adipocytes

FIGURE 6.5

LM of decalcified compact bone. The lamellae in the outer perimeter are bordered by periosteum and the bone marrow cavity. The bone marrow in this case has large amounts of adipose tissue surrounded by hemopoietic tissue. Owing to a high concentration of collagen fibers, the bony matrix in decalcified bone is highly eosinophilic. (100×)

FIGURE 6.6

LM of decalcified compact bone at a higher magnification. The lamellae on the outer perimeter are bordered on one side by the periosteum. The section shows numerous Haversian systems with osseous lamellae concentrically arranged around a centrally located Haversian canal. Unlike the chondrocytes in cartilage, osteocytes generally do not completely fill the lacunae in bone matrix. Several collapsed blood vessels and osteoblast cells bordering the bony matrix can be easily identified in this micrograph. (200×)

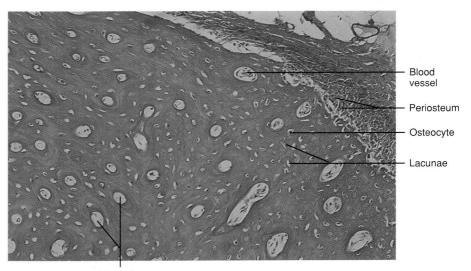

Blood vessel

Periosteum

Osteocyte

Lacunae

Haversian canals

FIGURE 6.7

LM of intramembranous trabeculae lined by bone-depositing osteoblasts. Osteocytes can be seen in their lacunae in the trabeculae. Hemopoietic tissue fills the cavities. (100×)

Periosteum

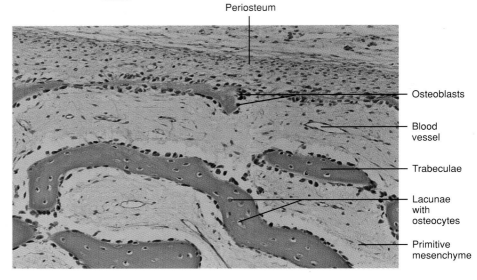

Osteoblasts

Blood vessel

Trabeculae

Lacunae with osteocytes

Primitive mesenchyme

Periosteum   Osteoclast   Lacunae

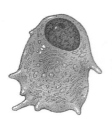

Osteoblast

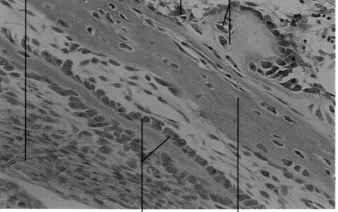

Osteoblasts   Trabeculae

FIGURE 6.8

LM of bone formation by osteoblast cells in intramembranous fetal bone. The micrograph shows osteoblasts depositing organic components that will become part of the bone matrix. Before mineralization occurs, the organic products are osteoids. The mineralization of osteoids takes place by the infiltration of calcium and phosphate ions, which lead to the formation of hydroxyapatite crystals between the associated collagen and ground substance. In contrast to mature osteocytes, osteoblast cells are large and have prominent nuclei and abundant basophilic cytoplasm. (400×)

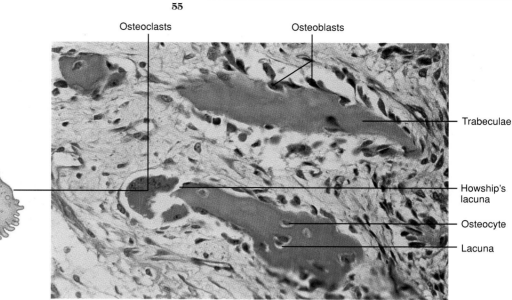

Osteoclast

FIGURE 6.9
LM of developing skull bone, displaying bone deposition by osteoblasts and bone reabsorption by osteoclast cells. Osteoclasts are large multinucleated cells formed by the cohesion of many osteoprogenitor cells. Bone reabsorption activity of the osteoclasts leads to the formation of bone depressions called Howship's lacunae. (400×)

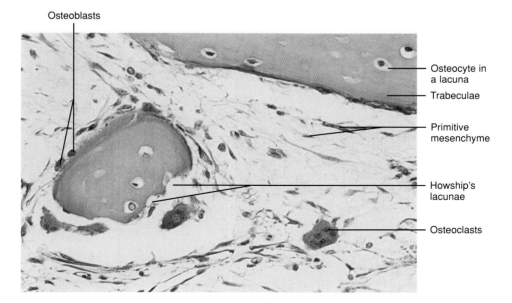

FIGURE 6.10
LM of membranous trabeculae from a human fetal skull. The trabecular surface is covered by osteoid-producing osteoblasts. Multinucleated giant osteoclast cells enzymatically bring about bone reabsorption, forming bone depressions called Howship's lacunae. Surrounding the trabeculae are collagen fibers and hemopoietic tissue. (400×)

FIGURE 6.11

LM of the border zone between epiphyseal cartilage and the diaphysis of a developing intercartilaginous long bone. Toward the epiphysis, the marrow cavity is considerably enlarged as it merges with the hyaline cartilage. The junction areas show hypertrophy of cartilage cells and calcification of the matrix. Red marrow surrounds the developing bone and remnants of cartilage. The osteoprogenitor cells are differentiating into osteoblasts which form the bony matrix. Surrounding the diaphysis is the beginning of periosteal bone or bone collar. The perichondrium will soon be replaced by periosteum. The replacement of cartilage by bone will continue until the primary ossification center of the diaphysis fuses with the secondary ossification center of the epiphysis. This fusion will lead to the formation of an epiphyseal plate or ring. (40×)

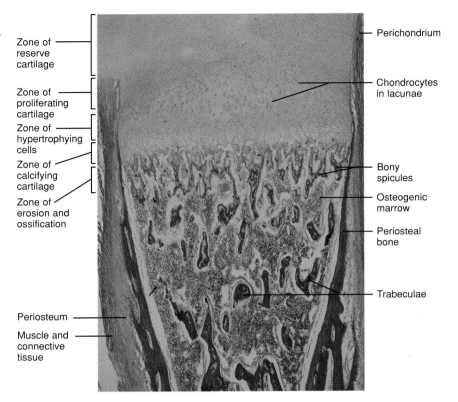

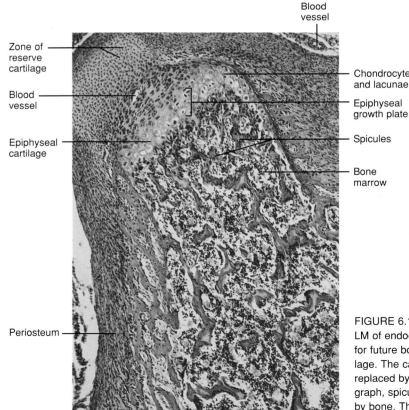

FIGURE 6.12

LM of endochondral ossification in a developing bone. The model for future bone is first established by a framework of hyaline cartilage. The cartilage is then gradually degraded by enzymes and replaced by bony matrix deposited by osteoblasts. In this micrograph, spicules with narrow areas of cartilage are being replaced by bone. The perichondrium has not differentiated into the periosteum and bone collar at this stage. (100×)

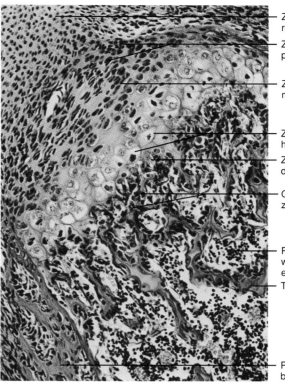

Zone of reserve cartilage

Zone of proliferation

Zone of maturation

Zone of hypertrophy

Zone of cartilage degeneration

Osteogenic zone

Primitve marrow with myeloid elements

Trabeculae

Periosteal bone

FIGURE 6.13
LM of endochondral ossification in a fetal bone at a higher magnification. Embryonic hyaline cartilage precedes the formation and deposition of bony matrix. In this micrograph, the following zones of intercartilaginous bone formation can be identified: zone of reserve cartilage, zone of proliferating cartilage, zone of hypertrophying cells and lacunae, zone of erosion and ossification, and zone of newly established periosteal bone. Also, bone marrow fills the upper medullary cavity of the diaphysis. (200×)

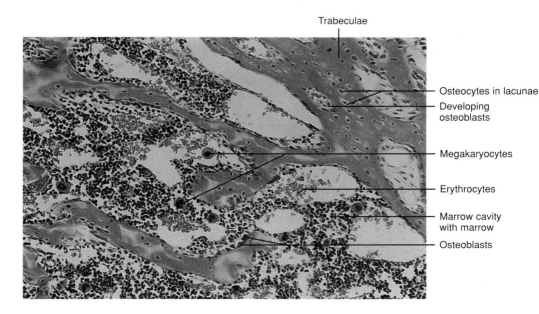

Trabeculae

Osteocytes in lacunae

Developing osteoblasts

Megakaryocytes

Erythrocytes

Marrow cavity with marrow

Osteoblasts

FIGURE 6.14
LM of cancellous bone, illustrating the nature of slender, trabeculae (which are also anastomosing), and irregular marrow cavities filled with hemopoietic tissue. The trabeculae exhibit lacunae with osteocytes, which can be seen all through their bony matrix. Several white blood cells, a concentration of red blood cells, and undifferentiated cells make up the bulk of the bone marrow. (200×)

FIGURE 6.15
LM of a developing intramembranous bone in a human
fetal cranium. The mesenchymal cells differentiate into
osteoblasts that secrete a noncalcified ground substance
called osteoid. Later, osteoblasts are gradually incorporated into the osteoid. At that stage, they are called
osteocytes. As calcium salts infiltrate the osteoid, it
hardens and forms the membrane bone. In the upper part
of the micrograph, the epidermis and dermis, with hair follicles, form a covering over the periosteum. Below the
periosteum, anastomosing trabeculae surround the
marrow cavities, which are filled with hemopoietic and
connective tissue. (40×)

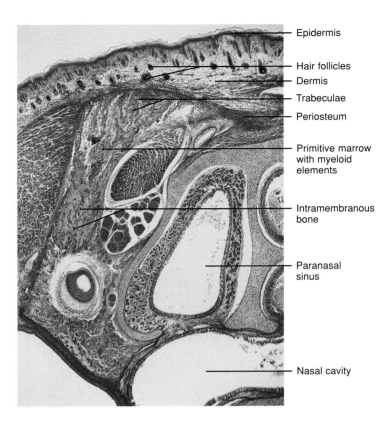

Epidermis

Hair follicles

Dermis

Trabeculae

Periosteum

Primitive marrow
with myeloid
elements

Intramembranous
bone

Paranasal
sinus

Nasal cavity

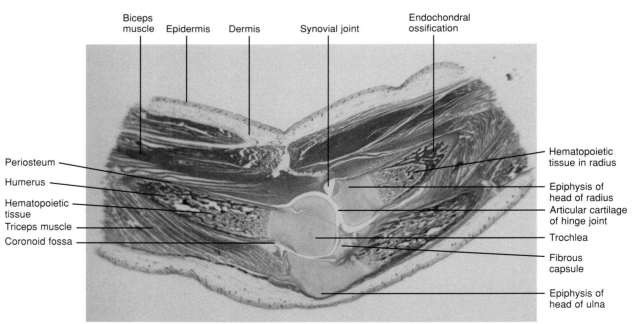

Biceps
muscle   Epidermis   Dermis   Synovial joint   Endochondral
ossification

Periosteum

Humerus

Hematopoietic
tissue

Triceps muscle

Coronoid fossa

Hematopoietic
tissue in radius

Epiphysis of
head of radius

Articular cartilage
of hinge joint

Trochlea

Fibrous
capsule

Epiphysis of
head of ulna

FIGURE 6.16
LM of a human fetal elbow joint, illustrating the developing long bones, muscles, connective
tissue, and skin. The cartilaginous epiphyseal ends of the three long bones—humerus, radius,
and ulna—form a synovial, or joint cavity. The diaphysis and the medullary cavities are at an
early stage of development. The cavities are filled with bony spicules and marrow. Surrounding the bones are narrow areas of metaphysis where the transition from cartilage to bone is
taking place. The synovial, or joint cavity, is encapsulated with a fibrous capsule, and synovial
folds can be seen between the articulating surfaces of the developing bones. (1×)

Parts of a Long Bone

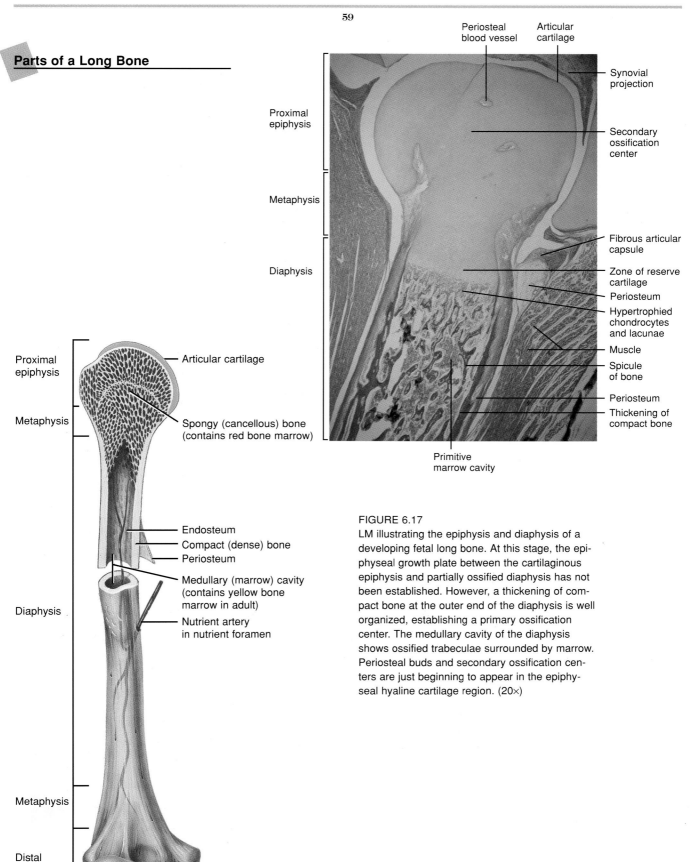

Periosteal blood vessel

Articular cartilage

Synovial projection

Secondary ossification center

Fibrous articular capsule

Zone of reserve cartilage

Periosteum

Hypertrophied chondrocytes and lacunae

Muscle

Spicule of bone

Periosteum

Thickening of compact bone

Proximal epiphysis

Metaphysis

Diaphysis

Primitive marrow cavity

Proximal epiphysis

Metaphysis

Diaphysis

Metaphysis

Distal epiphysis

Articular cartilage

Spongy (cancellous) bone (contains red bone marrow)

Endosteum

Compact (dense) bone

Periosteum

Medullary (marrow) cavity (contains yellow bone marrow in adult)

Nutrient artery in nutrient foramen

Articular cartilage

FIGURE 6.17
LM illustrating the epiphysis and diaphysis of a developing fetal long bone. At this stage, the epiphyseal growth plate between the cartilaginous epiphysis and partially ossified diaphysis has not been established. However, a thickening of compact bone at the outer end of the diaphysis is well organized, establishing a primary ossification center. The medullary cavity of the diaphysis shows ossified trabeculae surrounded by marrow. Periosteal buds and secondary ossification centers are just beginning to appear in the epiphyseal hyaline cartilage region. (20×)

Eponychium

Nail   Developing bone

Epidermis

Dermis

Distal
phalanx

Dense
Bone

Muscle
and connective
tissue

Periosteum

Synovial
projection

Articular
cartilage

Cartilage

Proximal
phalanx

Synovial
projection

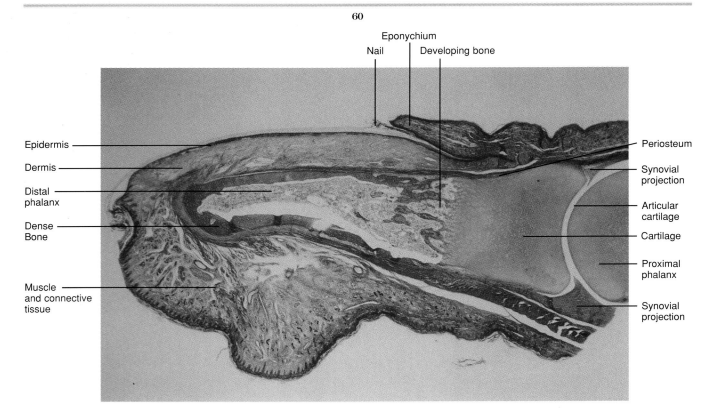

FIGURE 6.18

LM of a fetal finger, showing a fingernail, the synovial joint between the distal and medial phalanges, muscle,
connective tissue, and the epithelium. The distal phalanx shows well-established hyaline cartilage zones in the
epiphysis, and the proximal part of the diaphysis. The medullary cavity shows trabeculae and marrow and the
establishment of an outer dense bone lining, which becomes the primary ossification center. Also shown are
synovial projections between proximal and distal phalanges. (20×)

Specialized Connective Tissue: Blood

Blood is a specialized connective tissue in which millions of cells are suspended in a fluid medium called plasma. Blood has several diverse functions, including the transport of nutrients, gases, hormones, proteins, cellular waste products, lipids, and, of course, cells.

Blood plasma is an aqueous solution composed of inorganic salts, proteins (prothrombin, globulins, fibrinogen), and albumins. These proteins and electrolytes collectively create a colloidal osmotic pressure within the cardiovascular system that facilitates the regulation of aqueous solutions between extracellular (interstitial) fluid and plasma. In the plasma, albumin proteins bind to fatty acids, which are then transported to different parts of the body. In this way, albumin proteins serve as transport proteins for large molecules.

Globulins (antibodies) are complex proteins that are part of the lymphatic system and are involved in the immune process. Specific antibodies bind to specific antigens, forming an antigen-antibody complex that destroys the antigen.

Fibrinogen and **prothrombin** (proteins found in the blood) are involved in a series of chemical reactions that leads to the formation of a blood clot at the site of rupture in a blood vessel.

Blood cells are produced in the bone marrow by **hemopoiesis.** An exception to this rule are white blood cells (WBC, or leukocytes), which are formed in the bone marrow and the lymphatic system by extramedullary hemopoiesis. However, extramedullary hemopoiesis in a developing embryo can also occur in such places as the yolk sac, the liver, the spleen, and the lymph nodes. After birth, hemopoiesis is confined to the red marrow found in the long and flat bones of the body. In adults, hemopoiesis is confined to the flat bones.

Various types of blood cells can be identified in a sample of peripheral blood smear under a microscope. **Erythrocytes,** or red blood cells (RBC), make up over 99% of the blood cells. The remaining less than 1% of total blood cells is comprised of **leukocytes,** or white blood cells (WBC), and fragments of megakaryocyte cells called **platelets** or thrombocytes. These three groups of blood cells are classified, according to morphology and function, as follows.

1. Erythrocytes in mature form do not have nuclei. They stain with acid dyes because of the basic nature of hemoglobin. The cells are uniform in diameter, generally 7 to 8 microns, and are responsible for the transport of oxygen (O_2) and small amounts of carbon dioxide (CO_2).

2. Leukocytes, or white blood cells (WBC), are complete cells, each containing a nucleus and cell organelles. The WBC can be divided into two distinct groups based on the presence or absence of granules: granulocytes and nongranulocytes (also called agranulocytes). The granulocytes include eosinophils, neutrophils, and basophils. The agranular leukocytes include **monocytes** and **lymphocytes.** The agranulocytes contain some nonspecific granules, whereas the granulocytes always contain specific granules that are larger than the nonspecific granules. The three types of granular cells, and two types of agranular cells, have different functions.

3. Platelets, or **thrombocytes,** are fragments of megakaryocytes found in the marrow. These small disc-shaped fragments are approximately 2 to 4 microns in diameter and number from 200,000 to 350,000 per cubic millimeter of blood. Their specific function is associated with the clotting of blood.

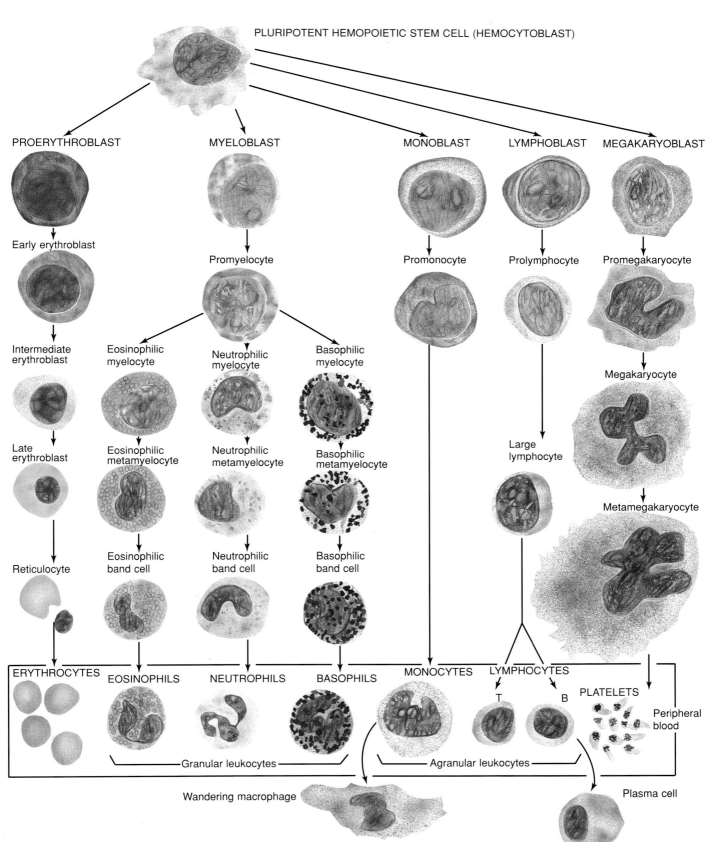

PLURIPOTENT HEMOPOIETIC STEM CELL (HEMOCYTOBLAST)

PROERYTHROBLAST   MYELOBLAST   MONOBLAST   LYMPHOBLAST   MEGAKARYOBLAST

Early erythroblast

Promyelocyte   Promonocyte   Prolymphocyte   Promegakaryocyte

Intermediate erythroblast   Eosinophilic myelocyte   Neutrophilic myelocyte   Basophilic myelocyte

Megakaryocyte

Late erythroblast   Eosinophilic metamyelocyte   Neutrophilic metamyelocyte   Basophilic metamyelocyte

Large lymphocyte

Reticulocyte   Eosinophilic band cell   Neutrophilic band cell   Basophilic band cell

Metamegakaryocyte

ERYTHROCYTES   EOSINOPHILS   NEUTROPHILS   BASOPHILS   MONOCYTES   LYMPHOCYTES   PLATELETS

T   B   Peripheral blood

Granular leukocytes   Agranular leukocytes

Wandering macrophage   Plasma cell

FIGURE 7.1

Diagrammatic representation of the genesis of blood cells. As shown, all blood cells are derived from pluripotent hemopoietic stem cells, the hemocytoblasts. Depending on the physiological needs of the body, the hemocytoblast cells go through several cycles of mitotic division and differentiation before maturing into erythrocytes, eosinophils, neutrophils, basophils, lymphocytes, monocytes, or megakaryocytes.

FIGURE 7.2

Light micrograph (LM) of erythrocytes in a peripheral blood smear. Erythrocytes (red blood cells) are highly specialized cells that are adapted to chemically bind and transport oxygen. The cells are biconcave, disc-shaped, pink, and 7 to 8 microns in diameter. They lack nuclei and are filled with hemoglobin. Like all blood cells, erythrocytes are derivatives of hemocytoblast stem cells. In the early stages of development, the erythrocytes are metabolically active and have nuclei, RNA, and the organelles necessary for protein synthesis. As the cells differentiate to form mature erythrocytes, the nuclei and organelles disintegrate, leaving only an outer plasma membrane that encloses the hemoglobin, a few necessary enzymes, and some hemoglobin protein. (1000×)

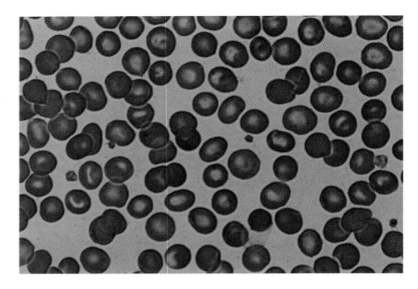

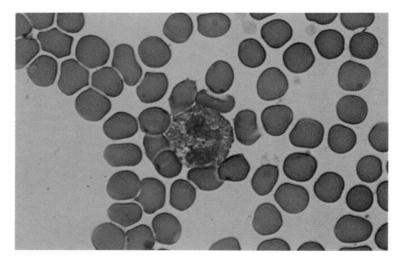

FIGURE 7.3

LM of an eosinophil in a peripheral blood smear. Eosinophils account for 1 to 6% of the leukocytes in circulating blood. They are 10 to 14 microns in diameter and contain large, refractile, spherical, reddish-orange granules. The nucleus is bilobed, and eosinophilic granules are of uniform size. Eosinophils are phagocytic cells that ingest microorganisms and antigen-antibody complexes. (1000×)

Barr body

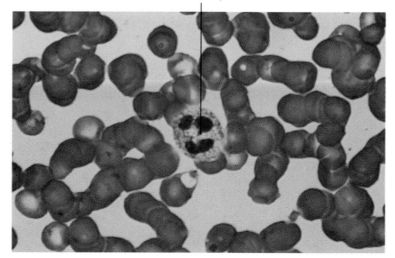

FIGURE 7.4

LM of a neutrophil in a smear of peripheral blood. Neutrophils are white blood cells with 3 to 5 multilobed nuclei. The cytoplasm contains small and large granules surrounded by a membrane (azrerophilic granules). The cells respond to acute inflammation and tissue damage. They are phagocytic cells that engulf cellular debris and microorganisms. In addition, they are highly mobile and can migrate from blood capillaries to sites of tissue damage. In a small proportion of females, the quiescent X chromosome forms a small discrete mass at the edge of the nucleus, called the Barr body. In the micrograph, the Barr body can be identified on one of the larger lobes of the nucleus. (1000×)

FIGURE 7.5
LM of neutrophils. The nuclei are multilobed. The cytoplasm contains large granules or primary granules, and small granules or secondary granules. Both primary and secondary granules are surrounded by membranes. The neutrophil granules contain phagocytins (antibacterial proteins) and the enzyme alkaline phosphatase. The other granules present in the cytoplasm are azurophilic lysosomes containing several hydrolytic enzymes, including myeloperoxidase, lysozyme, and D-amino-oxidase. (1000×)

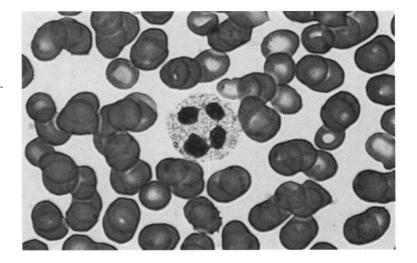

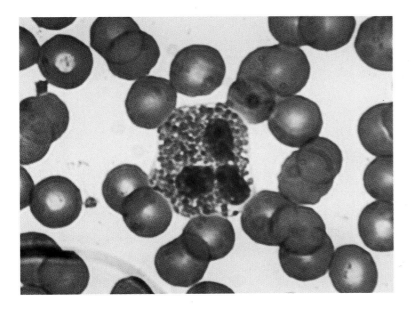

FIGURE 7.6
LM of a basophil in a peripheral blood smear. Basophils number less than 1% of the total circulating white blood cells (leukocytes). The nucleus is generally bilobed; however, most of the time it is obscured by large, densely basophilic granules. The granules are highly soluble in water and are generally dissolved during routine staining processes. This and the small number of cells add to the difficulty of identifying them. The granules are surrounded by a membrane and contain heparin (an anticoagulant), histamine (a vasodilator of small blood vessels), and a slow-reacting substance of anaphylaxis (SRS-A) that stimulates the contraction of smooth muscles in internal organs. The basophil cells are attracted to the sites of IgE antibody and antigen complexes that are formed during allergic reactions associated with allergic asthma and hay fever. (1000×)

FIGURE 7.7
LM of a monocyte in a smear of peripheral blood. The monocytes are the largest of the white blood cells and account for 2% to 10% of the total number of circulating leukocytes. The cell has a large, centrally placed nucleus, which is generally indented on one side, giving the nucleus a kidney shape. This is especially true of a mature cell. The cytoplasm contains small, nonspecific lysosomal granules. Monocytes are highly motile; they enter the connective tissue from the blood vascular system by diapedeses. In the tissue, they are known as tissue fixed cells, macrophages, or histiocytes. Macrophages are important in initiating an immune response and engulfing cellular debris. (1000×)

Monocyte   Platelet

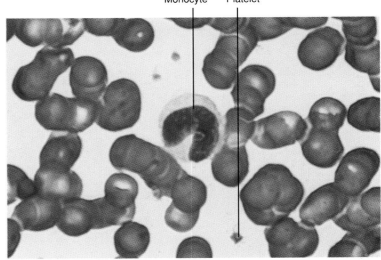

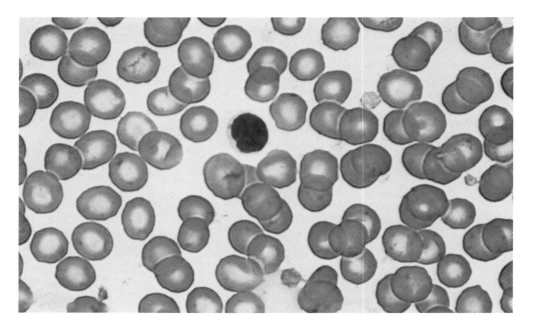

FIGURE 7.8

LM of a peripheral blood smear, illustrating a lymphocyte. Lymphocytes comprise 20% to 45% of the white blood cells. They are the smallest of all white cells, and are slightly larger than erythrocytes. These cells are characterized by large, round nuclei and small areas of pale, basophilic, nongranular cytoplasm. Lymphocytes play an important role in the immune process. (1000×)

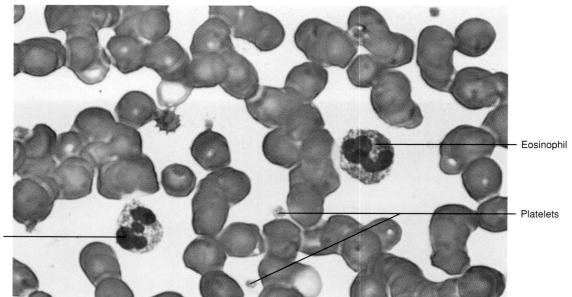

FIGURE 7.9

LM of platelets or thrombocytes. These are fragments of megakaryocyte cells found in the bone marrow. Mature, circulating platelets in humans contain two types of membrane-bound granules: dense granules, which store serotonin (5-hydroxytryptamine), Ca^{++}, ADP, and ATP; and alpha granules. The alpha granules contain the hydrolytic enzymes phosphatase and β-glucuronidase, platelet fibrinogen, platelet factor III, and thrombosthenin (a protein). (1000×)

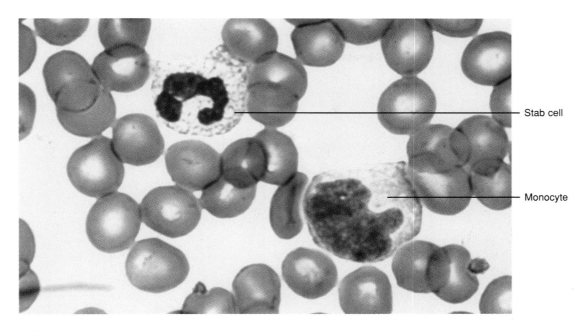

Stab cell

Monocyte

FIGURE 7.10
LM of a stab cell with a band-type nucleus, and a monocyte. Stab cells are immature neutrophils. The nucleus is segmented but has not reached the lobulation stage. (1000×)

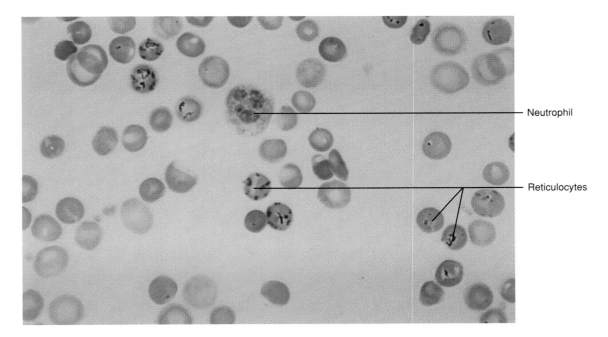

Neutrophil

Reticulocytes

FIGURE 7.11
LM of reticulocytes as seen in peripheral blood stained with crystal blue stain. Stained reticulocytes display a fine granular network of ribosomal RNA. Reticulocytes constitute less than 1% of the circulating erythrocytes and are slightly larger than the surrounding red blood cells, as seen in this micrograph. After severe erythrocyte depletion resulting from hemorrhage or certain pathologies, the reticulocyte count increases to a significant percentage, which is an indication of increased hemopoiesis in the bone marrow. (1000×)

Bone Marrow

In adult humans, hemopoiesis is confined to the bone marrow and the lymphoid tissue. However, in the embryonic and fetal stages, hemopoiesis occurs in the yolk sac, spleen, bone marrow, and liver. Under certain pathological conditions, the adult human spleen and liver may be stimulated to resume a role in hemopoiesis.

Hemopoiesis as discussed in this chapter is related to the active role the bone marrow plays in the generation of red and white blood cells and the platelets (thrombocytes). The genesis of blood cells is summarized in diagrammatic form in the previous chapter (Figure 7.1).

Bone marrow fills the medullary cavities of the cancellous, long, and intramembranous bones. Bone marrow cells are densely packed between blood vessels and reticular fiber stroma. Connective tissue cells—**osteoblasts, osteoclasts, plasma cells, macrophages, mast cells,** and bone marrow cells, which are multinucleated **megakaryocytes** and **adipocytes**—are also found in the hemopoietic tissue. Different sizes of blood vessels are also present in the marrow cavity.

The bone marrow has several functions: (1) the creation of blood cells and their migration into the peripheral blood, (2) the phagocytosis of defective cells, (3) the genesis of B and T lymphocytes, and (4) the formation of bone (**osteogenesis**) by the osteoblasts lining the endosteum.

Erythrocytes at different stages of development constitute a large proportion of the blood cells in the marrow. Easily recognizable are the **normoblasts** (with the darkly stained nuclei), **polychromatophil erythroblasts** (found individually, or in groups or nests), and **basophilic erythroblasts** (large cells that exhibit large, less dense, nuclei and a basophilic cytoplasm).

The granulocytes can be identified by the **polymorphonuclear heterophil granules,** which are analogous to granules of neutrophils and eosinophils. The precursors to the latter are **heterophilic myelocytes** which display large, round or ovoid nuclei. Less numerous and harder to recognize are the **reticular cells** and **hemocytoblast stem cells.**

Hematopoietic Tissue

FIGURE 8.1
Light micrograph (LM) of active bone marrow.
Present are numerous adipocytes, a meshwork of
collagen with reticular strands of connective tissue,
undifferentiated stem cells (progenitor cells for all cellular elements of blood), and vascular sinusoids,
which form narrow channels between regions of
hemopoietic tissue. As age progresses, the hemopoiesis in peripheral long bones slows down, and
gradually the hemopoietic tissue is replaced by inactive yellow adipose tissue. (100×)

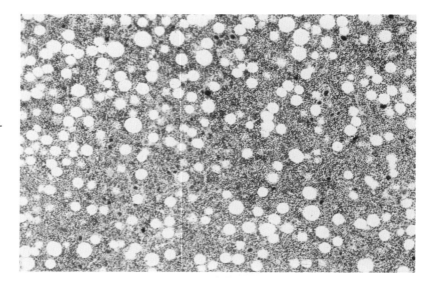

FIGURE 8.2
LM of active bone marrow at a higher
magnification. The micrograph shows
large concentrations of adipocytes,
differentiating stem cells, a few
megakaryocytes, and hemopoietic
cords that are separated by sinusoids
infiltrated by erythrocytes. (400×)

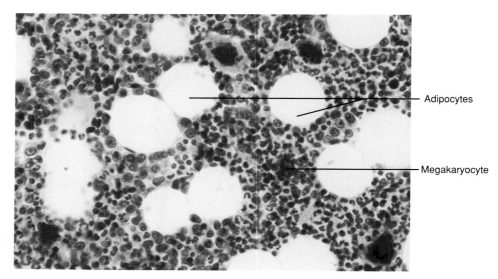

Adipocytes

Megakaryocyte

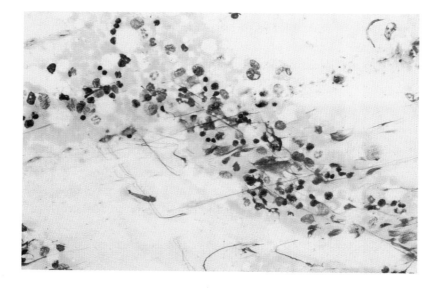

FIGURE 8.3
LM of active red bone marrow "touch" preparation
containing many cellular and streaked nuclear fragments. Typical mixed-cell hemopoietic cluster
extends diagonally across the field. (400×)

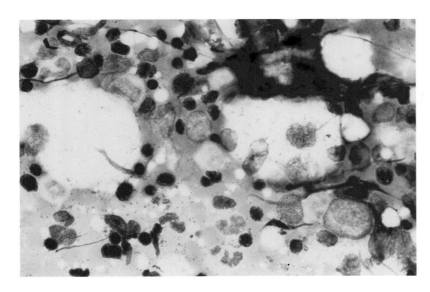

FIGURE 8.4
LM of active red bone marrow at a higher magnification, showing immature megakaryocytes and clustered myeloid and erythroid cell precursors, as well as scattered lymphocytes. Large numbers of adipocytes can also be identified in the micrograph. (1000×)

FIGURE 8.5
LM of the medullary cavity of a long bone, which is filled with hemopoietic tissue. Mature erythrocytes, differentiating megakaryocytes, and clusters of myeloid and erythroid cell precursors, as well as scattered lymphocytes, fill the cavities. Osteoblasts, osteocytes in lacunae associated with the trabaculeae, and osteoclasts can be identified in this micrograph. Also seen are trabeculae, with bone enzymatically reabsorbed by osteoclasts along their inner surfaces, and Howship's lacunae formed by this bone-reabsorption process. (400×)

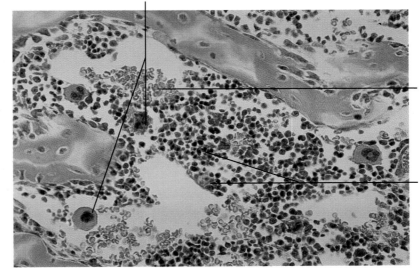

Megakaryocytes

Mature erythrocytes

Myeloid and erythroid precursors

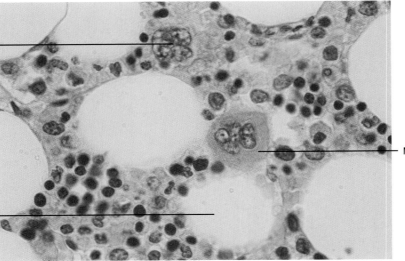

Megakaryoblast

Adipocyte

Megakaryocyte

FIGURE 8.6
LM of bone marrow section exhibiting stem cells that are maturing into blood cells. Large adipocytes, differentiated megakaryoblasts and megakaryocytes, can be identified in this micrograph. (1000×)

Muscle

uscle fibers can be classified into three groups based on location, morphology, and function: smooth muscle, skeletal muscle, and cardiac muscle fibers. All muscle fibers have distinctive functional similarities. However, there are slight differences in the morphologies of muscle fibers. The histological terminology of muscle tissue is slightly different from the terminology used in defining other tissues—e.g., the muscle plasma membrane is called sarcolemma, endoplasmic reticulum is called sarcoplasmic reticulum, and cytoplasm is called sarcoplasm.

Skeletal Muscle

The **skeletal muscle** is composed of voluntary fibers associated with the movement of the skeleton and certain organs, such as the tongue and the eye. The cross banding and alternating striation of actin and myosin is a characteristic feature of skeletal muscle. Under polarized light, anisotropic dark bands, called **A bands,** alternating with isotropic **I bands,** are visible. Running transversely in the center of the I band is a narrow dense band called the **Z band.** A

well-prepared muscle section under an electron microscope shows a narrow band, called the **H band,** passing through the middle of the A band. Dark **M bands** traverse the H band. Skeletal muscle fibers are multinucleate elongated cells surrounded by a membrane called the **sarcolemma.** The cells have a high concentration of golgi bodies, mitochondria, sarcoplasmic reticulum, and myofibrils. The myofibrils run parallel to the sarcoplasmic reticulum and form the myofilament striations of contractile actin and myosin proteins.

The relative thickness of the band can change depending on the state of contraction and the length of each fiber. The A band, however, does not change in length. The I band is short and narrow in a contracted muscle, but wide and prominent in a stretched muscle. A **sarcomere** is a muscle contractile unit that extends from one Z band to the next Z band. When the muscle fiber contracts, the I band becomes shorter and the Z bands approach the proximity of the A bands.

Each muscle fiber is surrounded by an external lamina, a sheath of reticular fibrils, and their surrounding fluids. This covering is called the

endomysium. Groups of a dozen or more muscle fibers surrounded by collagen and elastic fibers form a bundle of fibers called the **fasciculus.** In turn, the fasciculi are surrounded by an investment of collagen fibers mixed with elastic fibers. This external sheath of connective tissue is called the **perimysium.** Most skeletal muscles have many fasciculi. The fasciculi are surrounded by a dense sheath of connective tissue called the **epimysium.**

Cardiac Muscle

The **myocardium** or heart muscle forms a distinct middle layer that surrounds all the chambers of the heart. The myocardium has an external covering, the **epicardium,** and an internal covering, the **endocardium.**

The cardiac muscle fibers are composed of branching elongated cells, which indicate periodic contours where cells have formed junctions with adjoining cells. A characteristic feature of cardiac muscle is the presence of 0.5- to 1-nm-thick intercalated discs between cells. The discs are highly refractive in fresh muscle tissue and stain deeply in a fixed tissue. The single large oval nucleus is centrally located in each cell. Occasionally, there may be two nuclei in a given cell. However, this is quite rare.

The fibers, as in the case of skeletal muscle, contain actin and myosin myofilaments that are organized into **sarcomeres** in a manner similar to that seen in skeletal muscle. However, the myofibrils may branch and course more irregularly in cardiac muscle. Also, the myofilaments occasionally merge with adjacent myofibrils.

In mammalian cardiac tissue, each muscle fiber and its external lamina are surrounded by fine collagen and reticular fibrils. This covering is quite similar to the endomysium of skeletal muscle; however, it is not as well organized as the endomysium of skeletal muscle because of the apposing nature of each cardiac fiber.

Between the bundles of muscle fibers are coarser elastic and collagen fibers corresponding to the perimysium of the skeletal muscle. In the region of the atrioventricular junction, dense masses of connective tissue form the septa or separating zones.

Smooth Muscle

Like striated and cardiac muscle, **smooth muscle** is also derived from mesodermal mesenchymal cells. Exceptions are the iridic smooth muscle of the eye and the modified smooth muscle associated with the walls of sweat glands. These muscles are derived from the ectoderm.

Mature smooth muscle cells are **fusiform** or spindle-shaped and have flattened oval or elongated nuclei that are centrally located in abundant cytoplasm.

Densely packed muscle cells lie roughly parallel to each other in such a way that the wide portion of one cell nestles next to the narrow portion of the adjoining cell. Smooth muscles may differ in shape depending on their location: (1) short and thick in the walls of the small arteries; (2) twisted and folded, as a result of the contraction of elastic fibers, in the walls of large arteries; or (3) slender and very long in the body walls of the gastrointestinal tract.

The cytoplasm of smooth muscle cells contains long strands of contractile filaments, sarcoplasmic reticulum, golgi bodies, centrioles, ribosomes, glycogen, and, at times, lipid molecules. The filaments are aggregations of two contractile proteins, actin and myosin. However, in routine histological preparation, the actin filaments are much more clearly displayed than the myosin. To observe the myosin filaments, special fixation techniques are used to arrest the cell during contraction, which permits myosin or thick filaments to be identified in smooth muscle.

Smooth muscle can be found in many parts of the body, including the body wall region from the middle of the esophagus to the anus of the gastrointestinal tract, the hepatic ducts, the gallbladder, the dorsal wall of the trachea, the bronchial tree, the ureter, the urinary bladder, the urethra, the corpora cavernosa of the penis, the testes, the epididymis, the vas deferens, the seminal vesicle, the prostate and Cowper's glands, the broad ligament, the ovary, the oviduct, the uterus, the vagina, the blood vessels, the spleen and lymphatic ducts, the arrector pilorum, the iris, and the ciliary body of the eye.

Smooth Muscle

FIGURE 9.1

Light micrograph (LM) of visceral smooth muscle fibers in the body wall of the small intestine, and accompanying sketch. In this micrograph, muscle fibers are sectioned in transverse and longitudinal planes. Muscle fibers in a longitudinal plane exhibit compact, spindle-shaped fibers with well-defined, flat, long nuclei located in the center of the cells. The lower muscle layer shows fibers cut in a transverse plane. In some cases, the fibers show nuclei if the section has passed through the middle of the fibers. (100×)

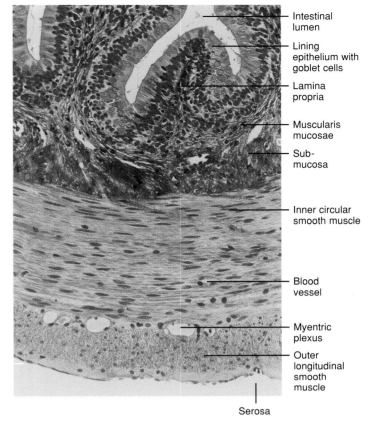

Intestinal lumen

Lining epithelium with goblet cells

Lamina propria

Muscularis mucosae

Sub-mucosa

Inner circular smooth muscle

Blood vessel

Myentric plexus

Outer longitudinal smooth muscle

Serosa

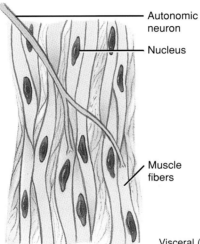

Autonomic neuron

Nucleus

Muscle fibers

Visceral (single-unit) smooth muscle tissue

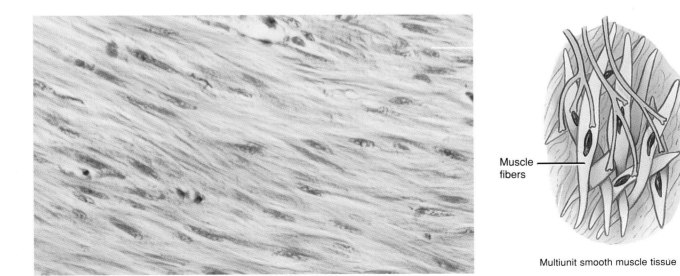

Muscle fibers

Multiunit smooth muscle tissue

FIGURE 9.2

LM of visceral smooth muscle at a higher magnification, and accompanying sketch. The fibers are spindle-shaped, elongated, and, at times, bifurcated. Generally, the smooth muscle cells are shorter than skeletal muscle fibers, and each cell has a single nucleus. The contractile proteins, actin and myosin, are not arranged into thick and thin filaments as in the skeletal and cardiac fibers, thus the name "smooth muscle." Between the fibers, a thin layer of connective tissue separates one fiber from another. The fibers of the connective tissue are analogous to the connective tissue that forms the endomysium of skeletal muscle. (400×)

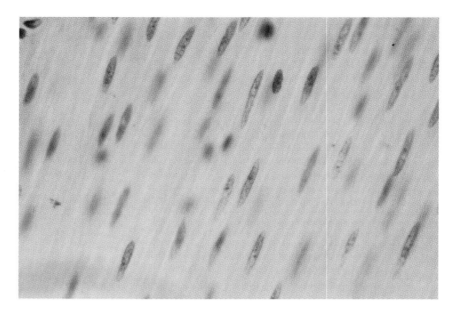

FIGURE 9.3
LM of visceral muscle fibers that have been cut longitudinally. The nuclei of the fibers are round, elongated, and centrally located. The cells are spindle-shaped with pointed ends. In this micrograph, the fibers are densely arranged. (400×)

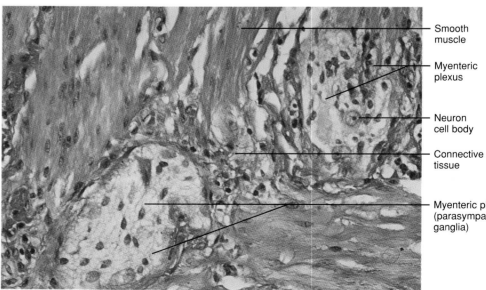

Smooth muscle

Myenteric plexus

Neuron cell body

Connective tissue

Myenteric plexus (parasympathetic ganglia)

FIGURE 9.4
LM of myenteric **(Auerbach's)** plexus in the body wall of the small intestine. The myenteric plexus is a concentration of parasympathetic ganglion cells and nerve fibers. The plexus is located in the connective tissue between the inner circular and outer longitudinal muscle layers of the muscularis externa. (400×)

Skeletal Muscle

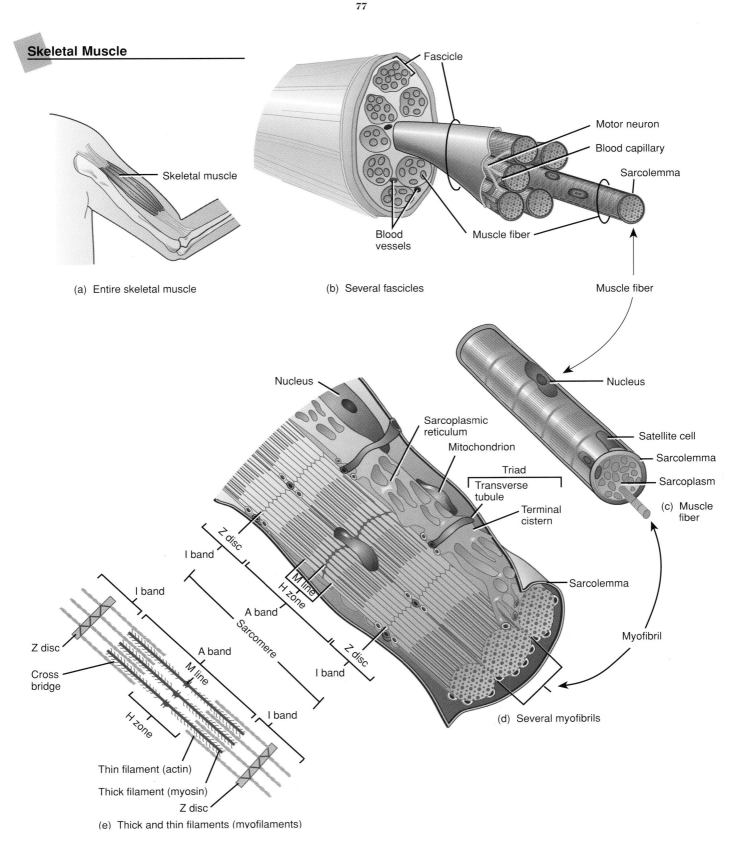

(a) Entire skeletal muscle

Skeletal muscle

(b) Several fascicles

Fascicle

Motor neuron

Blood capillary

Sarcolemma

Blood vessels

Muscle fiber

Muscle fiber

Nucleus

Satellite cell

Sarcolemma

Sarcoplasm

(c) Muscle fiber

Nucleus

Sarcoplasmic reticulum

Mitochondrion

Triad

Transverse tubule

Terminal cistern

Sarcolemma

Myofibril

(d) Several myofibrils

Z disc

I band

M line

H zone

A band

Sarcomere

A band

M line

Z disc

I band

I band

I band

Z disc

Cross bridge

H zone

Thin filament (actin)

Thick filament (myosin)

Z disc

(e) Thick and thin filaments (myofilaments)

FIGURE 9.5

Diagrammatic representation of skeletal muscle from gross to molecular level. Diagram (a) shows a muscle in the forearm region. Diagram (b) shows a cross section of muscle fascicles at a higher magnification. Diagram (c) shows a muscle myofibril. Diagram (d) shows several myofibrils in three dimensions to illustrate the internal morphology. Diagram (e) represents thin filaments (myofilaments) at still higher magnification.

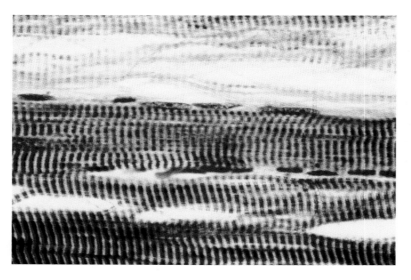

FIGURE 9.6
LM of skeletal muscle, demonstrating
the banding arrangement of myosin filaments, and accompanying sketch.
The actin filaments are not discernible
at this magnification. The elongated,
multinucleated muscle fibers show
numerous flattened nuclei located at
the peripheries of the cells, just
beneath the sarcolemma. The nuclei
of the fibroblast cells can be identified
between the fibers. (1000×)

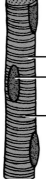

Skeletal muscle fiber

Nucleus of skeletal muscle fiber

Striation

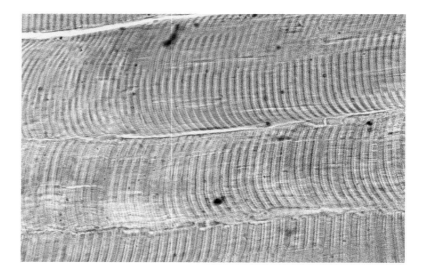

FIGURE 9.7
LM of skeletal muscle in a longitudinal section. The
muscle fibers illustrate the cross-banding arrangement
of actin and myosin filaments. The thick myosin
anisotropic A bands alternate with the thin actin
isotropic I bands. (1000×)

FIGURE 9.8
LM of skeletal muscle, demonstrating closely arranged
parallel myofibrils, which give the fibers a striated
appearance. Owing to a rupture of the sarcolemma, a
small portion of the fiber has separated from the main
strand. However, the A bands of the separated part of
the strand are still aligned with the A bands of the
myofibrils. (100×)

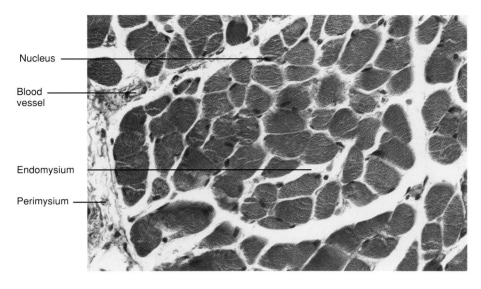

Nucleus

Blood vessel

Endomysium

Perimysium

FIGURE 9.9
LM of skeletal muscle fibers in cross (transverse) section, and sketch illustrating relationships between connective tissue and skeletal muscle (note relative positions of epimysium, perimysium, and endomysium). The peripheral location of the nuclei and the interfibrillar spaces are quite prominent. The connective tissue of the endomysium and cross sections of the nuclei of fibroblasts are identifiable between fibers. Connective tissue of perimysium surrounds the muscle fascicle. A small blood vessel can also be identified. (400×)

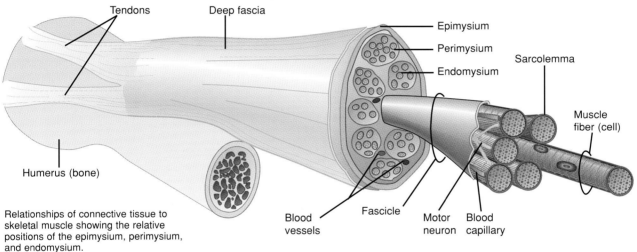

Tendons

Deep fascia

Epimysium

Perimysium

Endomysium

Sarcolemma

Muscle fiber (cell)

Humerus (bone)

Blood vessels

Fascicle

Motor neuron

Blood capillary

Relationships of connective tissue to skeletal muscle showing the relative positions of the epimysium, perimysium, and endomysium.

FIGURE 9.10
LM of a tendon/muscle junction in a longitudinal section. The micrograph displays skeletal muscle inserted into and surrounded by the tendon. Tendons are the densest form of fibrous connective tissue. The tissue contains bundles of coarse masses of collagen fibers and fibroblasts. (100×)

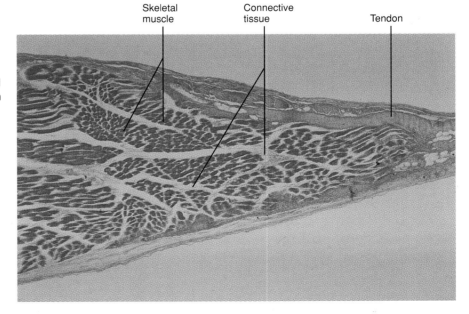

Skeletal muscle

Connective tissue

Tendon

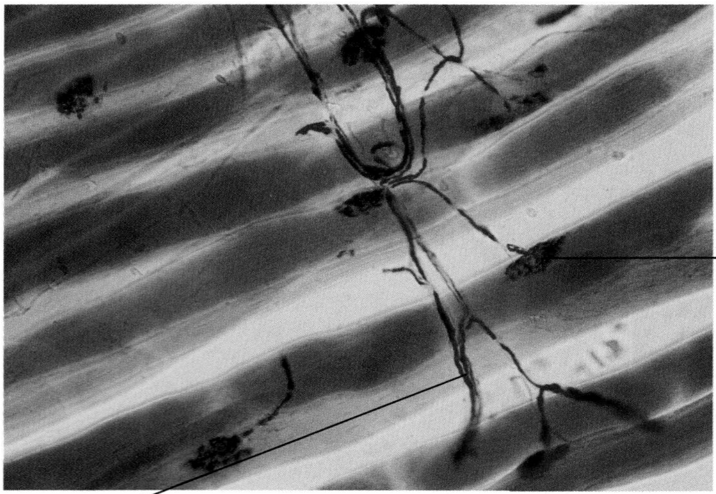

FIGURE 9.11
LM of neuromuscular junction (motor end plate), and accompanying sketch. In the micrograph, a branching axon of a motor neuron is innervating several muscle fibers. The axon terminal divides into several branches, each branch usually forming a neuromuscular plate on a different muscle fiber. (200×)

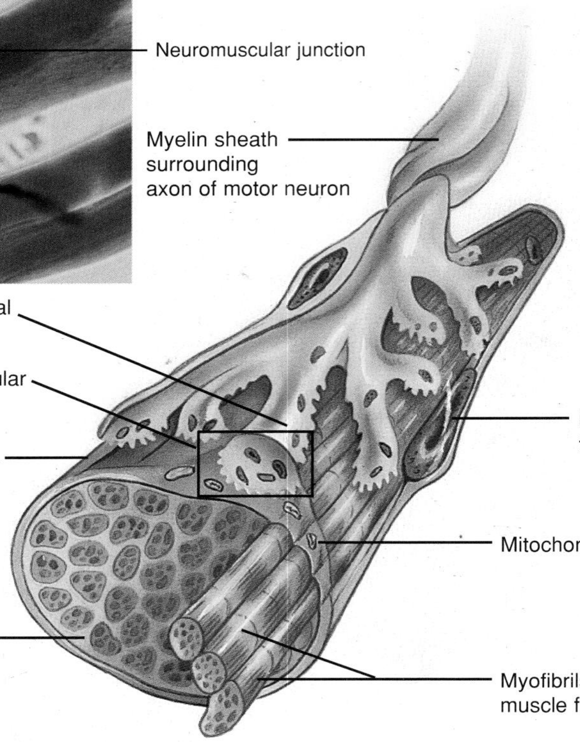

— Neuromuscular junction

Axon

Myelin sheath surrounding axon of motor neuron

Axon terminal

Neuromuscular junction

Sarcolemma

Nucleus of muscle fiber (cell)

Mitochondrion

Sarcoplasm

Myofibrils of muscle fiber

Cardiac Muscle

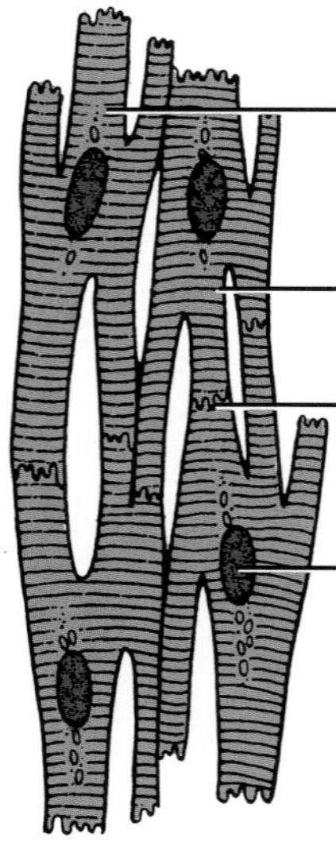

Cardiac muscle fiber

Striation

Intercalated disc

Nucleus of cardiac muscle fiber

FIGURE 9.12
LM of fetal myocardial tissue, and accompanying sketch. In this micrograph, the myocardial fibers have not fully matured. The fibers lack myofilaments, striations, and intercalated discs. However, the branching of the fibers, the large, oval, centrally placed nucleus in each cell, and the interfibrillar spaces, typical of cardiac fibers, are prominently displayed. (400×)

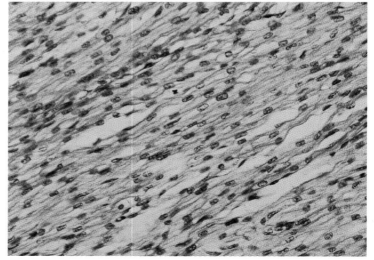

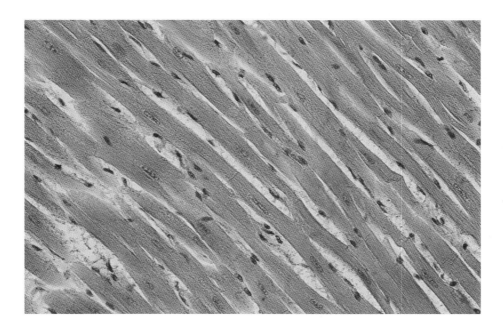

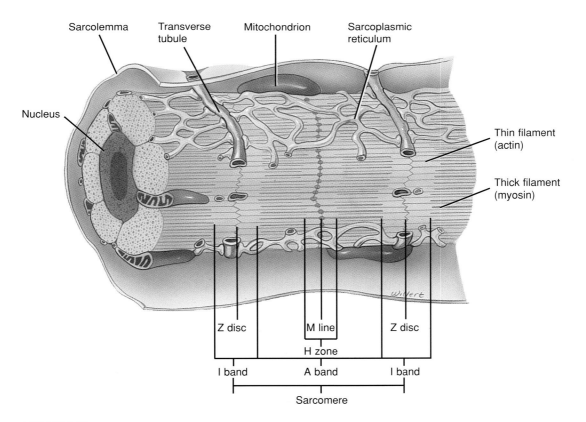

FIGURE 9.13

LM of cardiac muscle in a longitudinal section, and accompanying sketch. In the micrograph, cardiac fibers display morphological and functional characteristics that are intermediate between those of visceral and skeletal muscles. Generally, there is only one nucleus per cell; however, occasionally there may be two nuclei in a given cell. The nucleus is centrally located and the cells have striations as in skeletal muscles. Also, as depicted in this micrograph, the cells are separated by a fine network of connective tissue that surrounds the muscle fibers. (400×)

FIGURE 9.14

LM of a longitudinal section of myocardium, and accompanying sketch. In
the micrograph, the myocardial cells
exhibit cross and longitudinal striations
of myofibrils. Prominent intercalated
discs form interconnecting structures at
the junction of the cells. Also, centrally
placed nuclei are present in many of
the myocardial cells. (1000×)

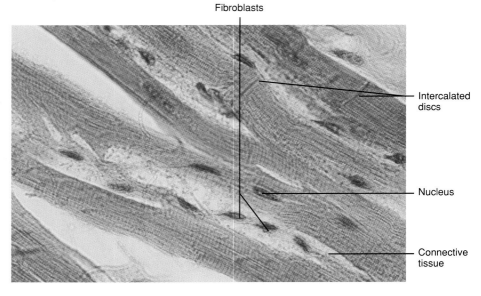

Fibroblasts

Intercalated
discs

Nucleus

Connective
tissue

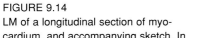

Cardiac muscle fibers

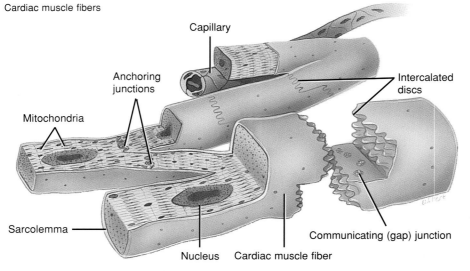

Capillary

Anchoring
junctions

Intercalated
discs

Mitochondria

Sarcolemma

Nucleus   Cardiac muscle fiber

Communicating (gap) junction

FIGURE 9.15

LM of a longitudinal section of
myocardium, displaying prominent oval
nuclei. At the poles of the nuclei, sarcoplasmic areas are relatively free from
myofibrils. Lipofuscin granules can be
seen in some of these areas. In addition, the myofibrils have shifted toward
the peripheries of the cells, forming
small groups of fibrils called
Cohnheim's fields. (1000×)

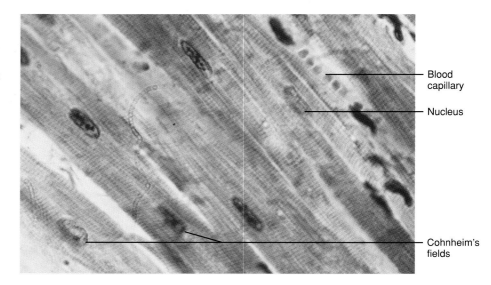

Blood
capillary

Nucleus

Cohnheim's
fields

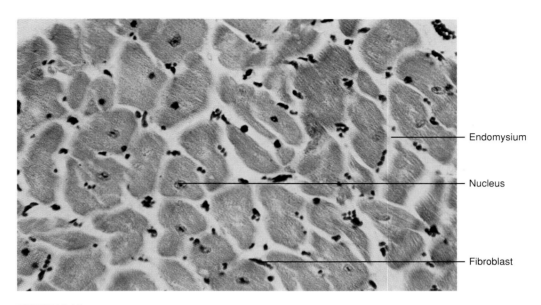

FIGURE 9.16
LM of cross section through myocardial tissue. The section passing through the center of the cell shows a prominent nucleus surrounded by a relatively myofibril-free area. (1000×)

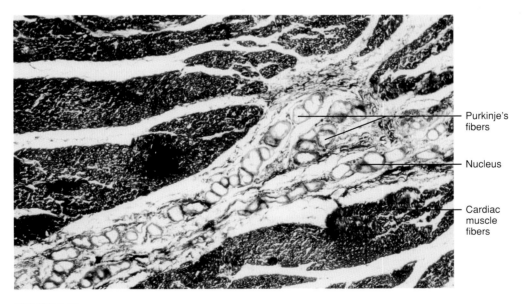

FIGURE 9.17
LM of **Purkinje's** fibers (cells) in longitudinal section. The Purkinje's fibers differ from cardiac muscle fibers in several respects. The Purkinje's fibers are much larger, they have fewer myofibrils that are peripherally located, the cells contain more sarcoplasm, some cells may have two nuclei, and the cell junctions lack intercalated discs. Purkinje's cells are part of the impulse-conducting system of the heart. (100×)

Nervous Tissue

Nervous tissue receives stimuli from both external and internal environments. The tissue is composed of two types of cells: **neuroglia** cells (also called **supporting** cells) and **neurons** or **nerve** cells. Neurons are highly specialized and can transmit impulses from one part of the body to another. The neuroglia cells are generally smaller than neurons and are scattered throughout the nervous tissue. Neuroglia cells are a conglomerate of cells: **ependymal** cells, which line the ventricles and the neural canal; **astrocytes,** which facilitate the transfer of molecules between blood capillaries and neural cells; **oligodendrocytes,** which myelinate fibers in the central nervous system; and **microglia,** which are phagocytic cells of the nervous system. Collectively, astrocytes and oligodendrocytes are referred to as **macroglia.**

The nervous system is divided morphologically into the central nervous system (**CNS**), which includes the brain and the spinal cord, and the peripheral nervous system (**PNS**), which includes all nervous tissue that is not part of the CNS. The PNS is nervous tissue that has specialized structures and connects all organ systems of the body to the CNS. The PNS includes all **nerves.**

This chapter will examine nervous tissue on a cellular level and will include neurons, sensory receptors, ganglia, nerves, ependymal cells, and modified cellular structures such as the choroid plexus found in the ventricles of the CNS. The CNS, which includes the brain and the spinal cord, will be discussed as part of the organ system in Chapter 11.

Neurons

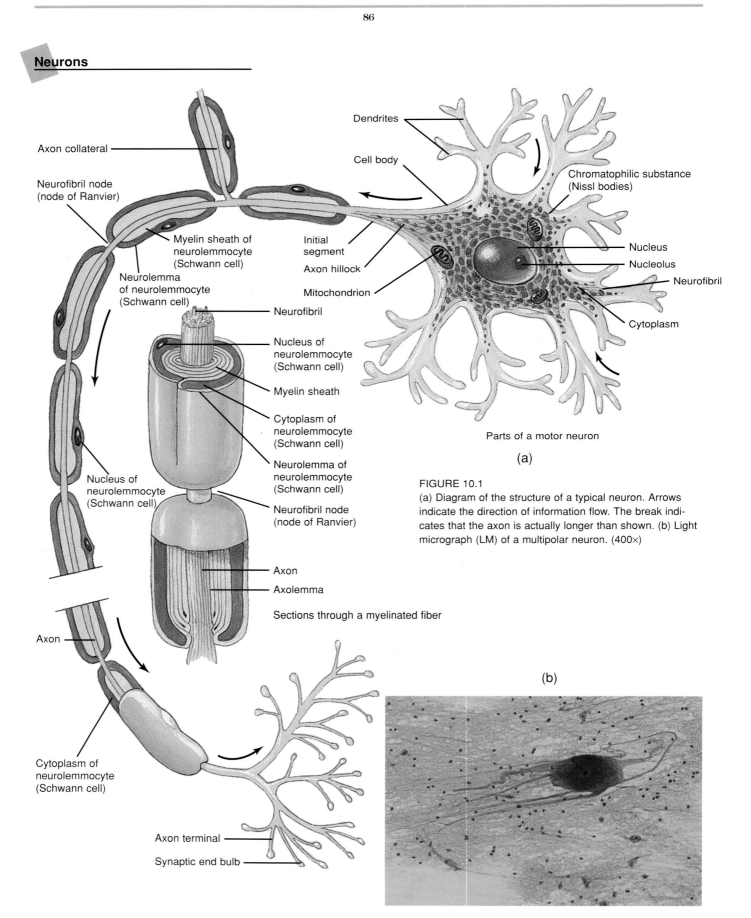

Axon collateral

Neurofibril node
(node of Ranvier)

Myelin sheath of
neurolemmocyte
(Schwann cell)

Neurolemma
of neurolemmocyte
(Schwann cell)

Neurofibril

Nucleus of
neurolemmocyte
(Schwann cell)

Myelin sheath

Cytoplasm of
neurolemmocyte
(Schwann cell)

Neurolemma of
neurolemmocyte
(Schwann cell)

Neurofibril node
(node of Ranvier)

Nucleus of
neurolemmocyte
(Schwann cell)

Axon

Axon

Axolemma

Cytoplasm of
neurolemmocyte
(Schwann cell)

Axon terminal

Synaptic end bulb

Sections through a myelinated fiber

Dendrites

Cell body

Chromatophilic substance
(Nissl bodies)

Initial
segment

Axon hillock

Mitochondrion

Nucleus

Nucleolus

Neurofibril

Cytoplasm

Parts of a motor neuron

(a)

FIGURE 10.1
(a) Diagram of the structure of a typical neuron. Arrows
indicate the direction of information flow. The break indicates that the axon is actually longer than shown. (b) Light
micrograph (LM) of a multipolar neuron. (400×)

(b)

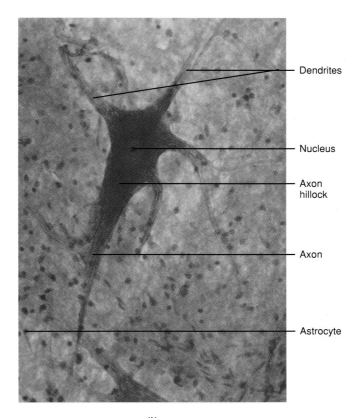

(i)

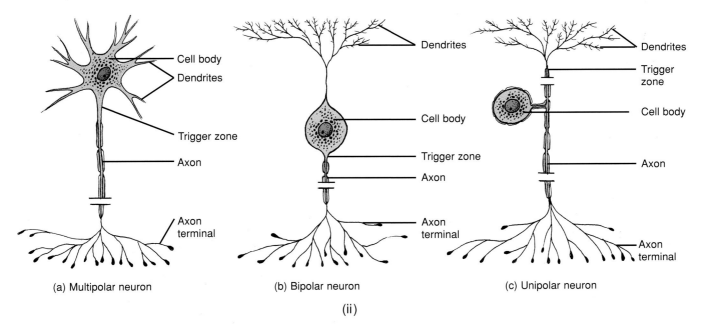

(a) Multipolar neuron   (b) Bipolar neuron   (c) Unipolar neuron

(ii)

FIGURE 10.2

(i) LM of a giant multipolar neuron, illustrating the basic structures common to all neurons: a large cell body (perikaryon), and two types of processes—a single axon and several dendrites extending from the perikaryon. The axon extends as a cylindrical process away from the cell body and terminates in small branches that form junctions, or synapses, with effector organs. Depending on the type of neuron, the axon may be long (**Golgi Type I**) or short (**Golgi Type II**). The terminal ends of the axons form a small swelling called **boutons de passage.**

(ii) Structural classification of neurons based on number of processes extending from the cell body. (400×)

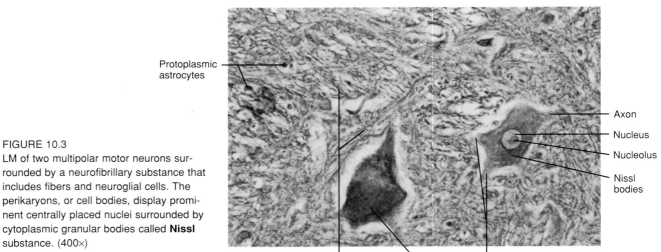

Protoplasmic astrocytes

Axon

Nucleus

Nucleolus

Nissl bodies

FIGURE 10.3
LM of two multipolar motor neurons surrounded by a neurofibrillary substance that
includes fibers and neuroglial cells. The
perikaryons, or cell bodies, display prominent centrally placed nuclei surrounded by
cytoplasmic granular bodies called **Nissl**
substance. (400×)

Fibers

Perikaryon
of a motor
neuron

Dendrites

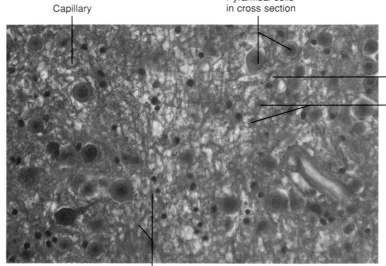

Capillary

Pyramidal cells
in cross section

Blood
capillary

Neuroglia
cells

Intercellular nerve
fibers and neuroglia

FIGURE 10.4
LM of giant pyramidal cells, in cross section, as seen in
the cerebral cortex. The cell body of each neuron exhibits
a large vesicular nucleus. Large numbers of neuroglia
cells, nerve fibers of various cells, and blood capillaries
can be identified in the micrograph. (200×)

Dendrites

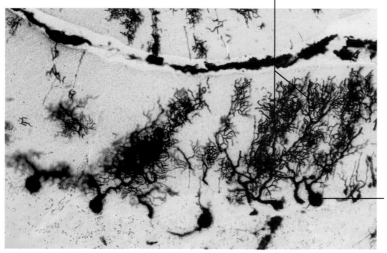

Cell
body

FIGURE 10.5
LM of silver stained **Purkinje** cells. Purkinje cells are neurons found in the stratum gangliosum (central layer) of the
cerebellar cortex. The Purkinje cell axon originates at the
lower pole of the cell body or perikaryon and extends
toward the granular layer. From there it continues to the
white matter of the cerebellum. The dendrites form fan-
shaped concentrations that reach up to the cerebellar surface. The bodies of Purkinje cells are surrounded by a
basket-like network of delicate nerve fibers of various origins. (400×)

Neuroglia Cells

FIGURE 10.6

LM of **Cajal** impregnated central nervous tissue displaying star-shaped branching fibrous astrocytes. The nuclei are large and oval or spherical in shape. The cytoplasmic processes extend laterally and abut with the blood vessels, forming perivascular feet. Other cytoplasmic extensions may connect to the surfaces of the neurons and the basement laminae of the blood capillaries. Protoplasmic astrocytes are found in the gray matter of the brain and spinal cord. Perivascular feet of some of the astrocytes are abutting with the blood vessels and the neurons. The astrocytes are surrounded by neuropil (fibrous) substance. (400×)

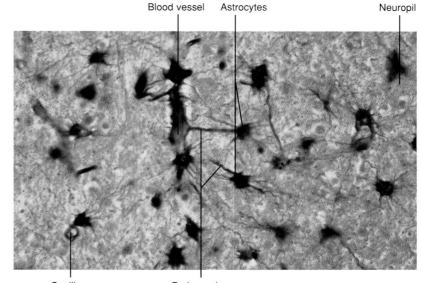

Blood vessel   Astrocytes   Neuropil

Capillary   Perivascular feet

Perivascular feet

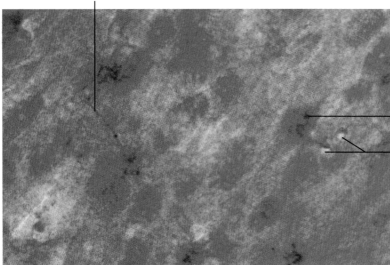

FIGURE 10.7

LM of fibrous astrocytes in the gray matter of the cerebral cortex. The cytoplasmic branching processes of some of the astrocytes are in contact with the blood capillaries. The astrocytes are surrounded by neuropil substance. (1000×)

— Astrocyte

— Blood capillaries

Protoplasmic astrocyte

FIGURE 10.8

LM of a protoplasmic astrocyte stained by a silver impregnation technique. Protoplasmic astrocytes are located in the gray matter of the brain and the spinal cord, and have many branching cytoplasmic processes. The perivascular feet are in contact with blood vessels and neuronal surfaces. In some regions of the nervous tissue, protoplasmic astrocytes connect to one another by forming gap junctions. (400×)

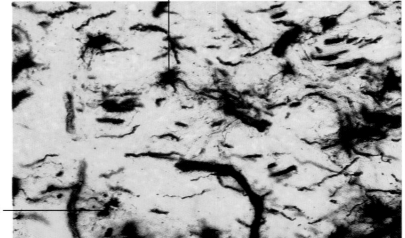

Microglia —

FIGURE 10.9
LM of oligodendrocyte (oligoden-
droglia) in the central nervous
system. Oligodendrocytes are located
in rows around blood vessels
(perivascular), between myelinated
fibers, adjacent to nerve cell bodies
(perineuronal), or in association with
the satellite cells. The perineuronal
cytoplasmic processes extend toward
the nerve fiber, wrap around it, and
form the myelin sheath. Oligodendrocytes are similar to the Schwann cells
of the PNS in function. (400×)

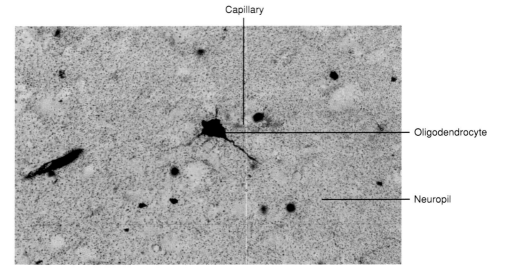

Capillary

Oligodendrocyte

Neuropil

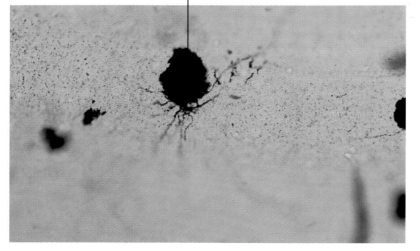

Microglia cell

FIGURE 10.10
LM of a silver stained microglia cell. The microglia
cells are of mesodermal origin and have a similar
phagocytic function as the macrophage of other tissues. The nucleus in microglia is elongated and
small, the cytoplasm is scanty, and the cells demonstrate numerous small spiny processes. Microglial
cells are usually found close to the blood vessels
throughout the gray and the white matter of the
CNS. (400×)

FIGURE 10.11
LM of ependymal cells that line the inner lining of the
ventricles and the central canal of the spinal cord.
The cells are cuboidal in shape and lack a basement
membrane. Cilia project from the cells into the luminal
space. The cilia in conjunction with the microvilli may
be involved in movement of the cerebrospinal fluid
within the cavities. (1000×)

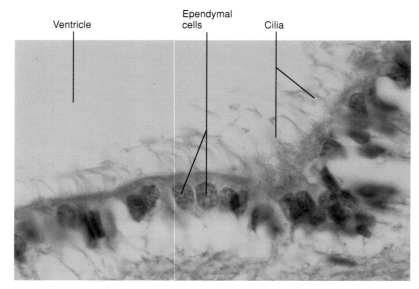

Ventricle

Ependymal
cells

Cilia

Nerves

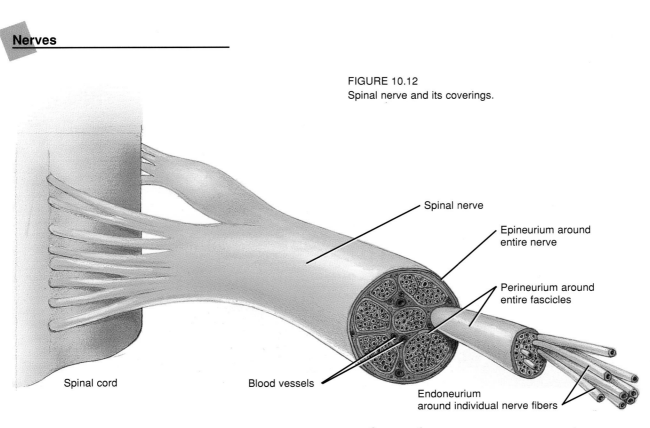

FIGURE 10.12
Spinal nerve and its coverings.

Spinal nerve

Epineurium around entire nerve

Perineurium around entire fascicles

Spinal cord

Blood vessels

Endoneurium around individual nerve fibers

Cross section

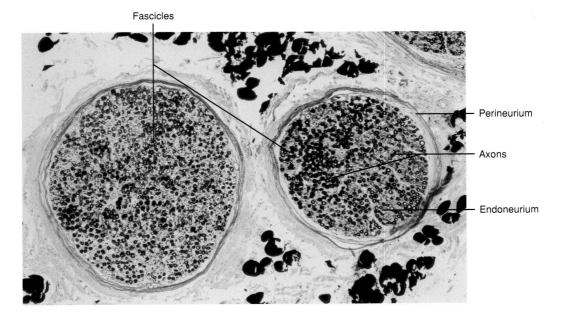

Fascicles

Perineurium

Axons

Endoneurium

FIGURE 10.13
LM of a cross section of two bundles (**fascicles**) of peripheral nerves with myelinated axon fibers. The nerves are surrounded by a covering of collagen fibers called perineurium. Connective tissue and blood vessels fill the spaces between the two nerves. The specimen has been stained with osmium tetroxide to show the myelin sheath covering of the axon fibers. (100×)

FIGURE 10.14
LM of a small peripheral nerve, in cross section, that has been stained with osmium tetroxide in order to show the myelin sheath around the axon fibers (nerve fibers). The nerve is surrounded by collagen fibers (perineurium). Individual nerve fibers are surrounded by connective tissue forming the endoneurium. **Schwann** cells, fibroblasts, and blood vessels can be identified between the axons. (200×)

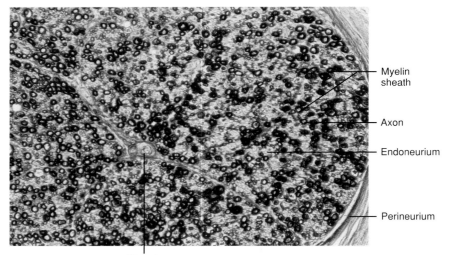

Myelin sheath

Axon

Endoneurium

Perineurium

Vessel

FIGURE 10.15
LM of two small (H and E stained) peripheral nerves in cross section with the neurovascular bundle. The staining method was used here to exhibit the axon fibers rather than the myelin sheath. The clear areas around the axon represent the myelin. The two nerve bundles are surrounded by the perineurium, which in turn is surrounded by the outermost sheath, the epineurium. The blood vessels, the collagen fibers, and the nuclei of fibroblasts can be seen in both the perineurium and epineurium. (400×)

Vessel   Capillary

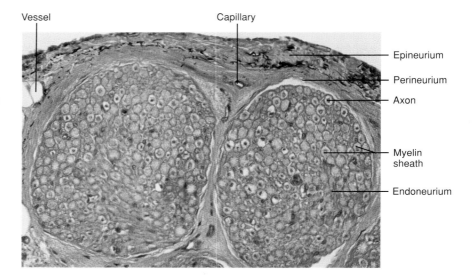

Epineurium

Perineurium

Axon

Myelin sheath

Endoneurium

Perineurium   Axon   Myelin sheath

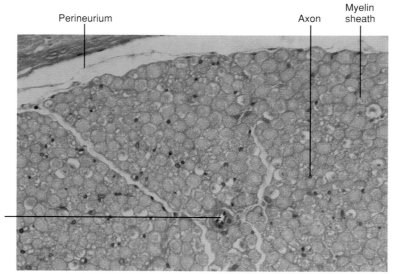

Vessel

FIGURE 10.16
LM of an H and E stained cross section of a dorsal root of the spinal nerve. The staining technique used illustrates the axon fibers instead of the myelin sheath. The variation in axon fiber surroundings is a result of the fibers being myelinated or nonmyelinated. The smaller fibers lack myelin, whereas the larger fibers show a precise concentration of myelin around the centrally placed axon. Crescent-shaped dark nuclei indicate the presence of **Schwann** cells, which elaborate myelin. Also seen in the micrograph are blood vessels, perineurium, and endoneurium. The clear area partition is an artifact of fixation. Dorsal root nerve fibers have a sensory function, whereas ventral root fibers have a motor function. (400×)

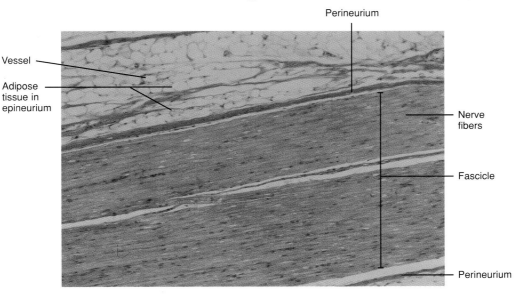

Vessel

Adipose tissue in epineurium

Perineurium

Nerve fibers

Fascicle

Perineurium

FIGURE 10.17

LM at low magnification of an H and E stained longitudinal section through a small portion of a peripheral nerve. The nerve fascicle is invested by the epineurium connective tissue with adipose cells on one side and by connective tissue of the perineurium on the other side. The surrounding tissue is composed of numerous nuclei of fibroblasts, **Schwann** cells, endothelial cells of blood capillaries, nerve fibers, and connective tissue fibers. At this magnification it is difficult to differentiate between Schwann cells and **fibroblast** cell nuclei. (100×)

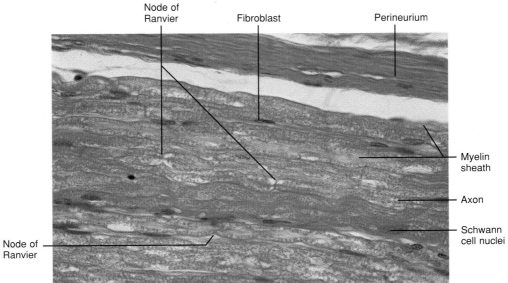

Node of Ranvier

Fibroblast

Perineurium

Myelin sheath

Axon

Schwann cell nuclei

Node of Ranvier

FIGURE 10.18

LM of a small portion of peripheral **nerve** in a longitudinal section, stained with H and E technique. In the micrograph, **Schwann** cells, fibroblast nuclei, nodes of **Ranvier,** and the endoneurium connective tissue between fibers can be identified. (400×)

Sensory Receptors

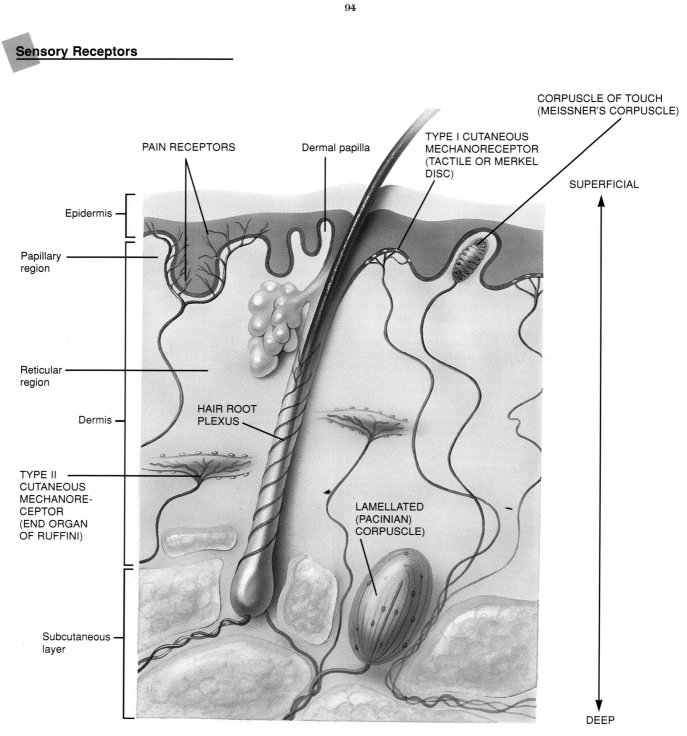

PAIN RECEPTORS

Dermal papilla

TYPE I CUTANEOUS MECHANORECEPTOR (TACTILE OR MERKEL DISC)

CORPUSCLE OF TOUCH (MEISSNER'S CORPUSCLE)

SUPERFICIAL

Epidermis

Papillary region

Reticular region

Dermis

HAIR ROOT PLEXUS

TYPE II CUTANEOUS MECHANORE- CEPTOR (END ORGAN OF RUFFINI)

LAMELLATED (PACINIAN) CORPUSCLE)

Subcutaneous layer

DEEP

FIGURE 10.19
Structure and location of cutaneous receptors.

Pacinian
corpuscles

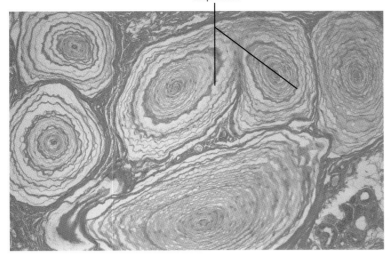

FIGURE 10.20
LM of a cross section of thick skin (dermis and subcutaneous tissue). Located in the dermis and the subcutaneous tissue are sensory receptors referred to as **Pacinian** corpuscles. The corpuscles are receptors for pressure and are possibly sensors of vibrations. In the micrograph, several corpuscles are cut transversely. A typical corpuscle is ovoid in shape, with an inner bulb located in the elongated central core. In a fresh specimen, the corpuscle has terminal myelinated nerve fibers surrounded by collagenous concentric lamellae. A dense connective tissue sheath invests the corpuscle. In the cutaneous and subcutaneous areas, the corpuscles are surrounded by connective tissue, blood vessels, sweat glands, and nerves. Pacinian corpuscles are also found in the pancreas (cat) and the body wall of the penis. Their function in the pancreas is uncertain. (100×)

Dermal   Meissner's
papilla   corpuscle   Epidermis

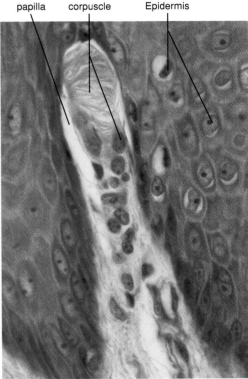

FIGURE 10.21
LM of **Meissner's** corpuscle in the dermis of the skin. These corpuscles are more prevalent in the integuments of fingertips, lips, and soles of the feet. They function as sensory receptors for discriminating light touch. The sensitivity of touch depends on the concentration of the corpuscles in a given area. Morphologically, the corpuscles are large, plump, and oval in shape. Nonmyelinated branches of the myelinated sensory fibers traverse the corpuscle. The corpuscles lie in the recesses of the dermal papillae. (1000×)

Motor   Terminal
Synapse   nerve fibers   fibril

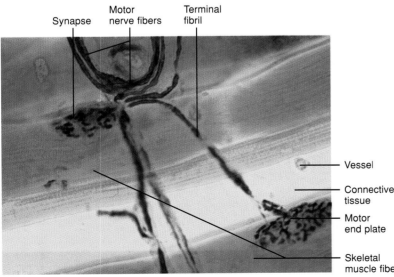

— Vessel

— Connective
tissue
— Motor
end plate

— Skeletal
muscle fibers

FIGURE 10.22
LM of a neuromuscular junction (neuromuscular plate). The distal end of a nerve fiber (axon) branches and distributes its branches over the muscle membrane (sarcolemma), establishing a specialized junction called a motor end plate. Identifiable in the micrograph are two well-established neuromuscular plates along with the axon fibers leading to the plates. (1000×)

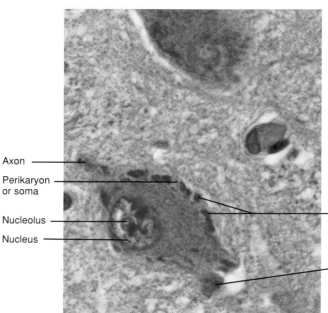

FIGURE 10.23

LM of synaptic knobs (boutons), which are the terminal portions of the axons, forming synapses with the soma or cell body of the neuron. The synaptic terminals shown in the micrograph are only a fraction of the terminals associated with the cell body, axon, and dendrites of a neuron. There may be as many as 1200 to 1800 boutons in contact with a single spinal motor neuron. (1000×)

Axon

Perikaryon or soma

Nucleolus

Nucleus

Axosomatic synapse

Dendrite

Ganglia

Spinal root nerve fibers   Fascicle of nerve fibers   Neurons

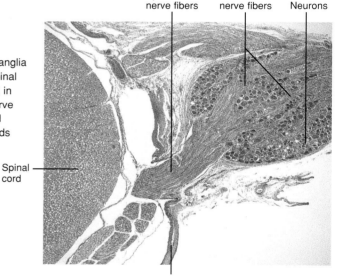

FIGURE 10.24

LM of a spinal ganglion in the dorsal root of the spinal nerve. Similar ganglia are associated with some of the cranial nerves and are called craniospinal ganglia. The neurons in the ganglion are pseudounipolar and spherical in shape. The micrograph shows nerve cell bodies between groups of nerve and connective tissue fibers. Parts of the spinal nerve, spinal cord, and meninges are also visible in the micrograph. A fibrous capsule surrounds the ganglia. (40×)

Spinal cord

Meninges

Ganglion cell bodies

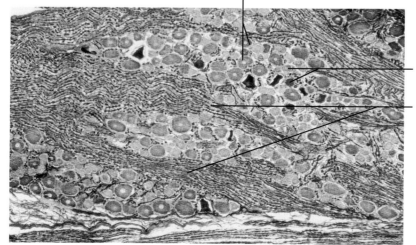

Satellite cells

Nerve fibers

FIGURE 10.25

LM, at a higher magnification, of the spinal ganglia. Visible are pseudounipolar neuron cell bodies in cross section, myelinated fibers that divide into peripheral and central branches, tufts of connective tissue fibers mixed with neuronal fibers, satellite cells, and part of a fibrous capsule that surrounds the ganglia. (100×)

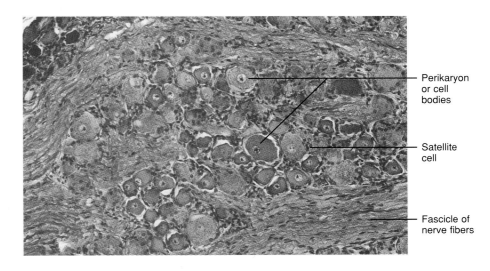

Perikaryon
or cell
bodies

Satellite
cell

Fascicle of
nerve fibers

FIGURE 10.26
LM of a spinal ganglion (dorsal root ganglion), displaying the perikaryons or cell bodies of the unipolar neurons surrounded by concentrically arranged satellite cells. Myelinated nerve fibers of the ganglion cells, blood capillaries, nuclei of fibroblasts, **Schwann** cells, and connective tissue fibers can also be identified. (400×)

Satellite cells

Cell bodies

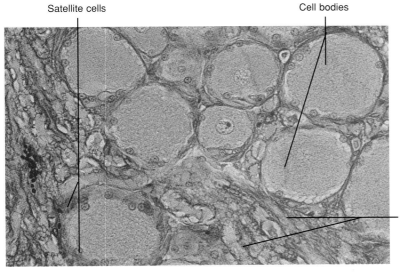

Nerve
fibers

FIGURE 10.27
LM of small portion of a dorsal root ganglion displaying various sizes of unipolar neuron cell bodies, which are ovoid or spherical in shape. Prominent large satellite cells are concentrically arranged around the peripheries of the cell bodies. **Schwann** cells, myelinated axons, dendrites, collagen and reticulate fibers, blood vessels, and fibroblasts surround the perikaryons. (400×)

FIGURE 10.28
LM of a cross section of sympathetic ganglia demonstrating structures similar to those seen in the spinal ganglia. However, the neurons in the sympathetic ganglia are widely placed, the cell bodies are multipolar, the nuclei are eccentrically located, and the perikaryons are separated by axons and dendritic, collagen, and reticulate fibers. Cellular debris in the form of lipofuscin granules can be seen in the nucleoplasm of some of the cells. (400×)

Multipolar
ganglion
cell bodies

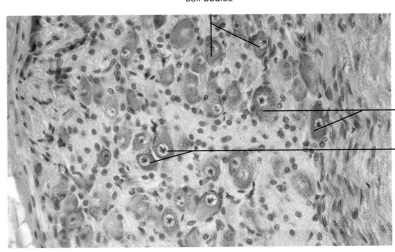

Lipofuscin
granules

Eccentrically
located
nuclei

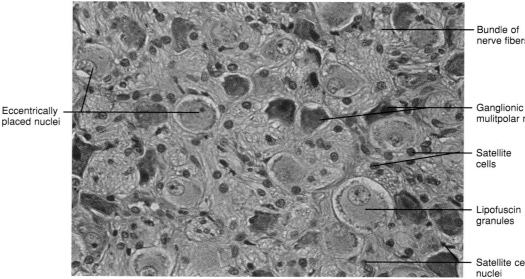

Eccentrically placed nuclei

Bundle of nerve fibers

Ganglionic mulitpolar neurons

Satellite cells

Lipofuscin granules

Satellite cell nuclei

FIGURE 10.29

LM of sympathetic ganglia. The micrograph shows several satellite cells, multipolar neurons with eccentrically placed nuclei in the cell bodies, and sparse connective tissue around the cell bodies. Preganglionic nerve fibers from the thoracolumbar outflow of the spinal cord synapse with the neurons in the sympathetic ganglia. The darkly stained nucleolus and **Nissl** substance can be identified in some of the perikaryons. Binucleated cell bodies are not uncommon, and lipofuscin pigment can be found in the cytoplasm of most of the neurons. There are fewer satellite cells surrounding the cell bodies of sympathetic ganglionic cells in comparison with the dorsal root ganglia. The surrounding fibers are a combination of collagen fibers of the connective tissue and preganglionic and postganglionic fibers associated with the spinal cord and the organs. (400×)

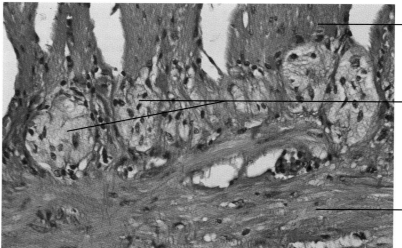

Muscularis

Parasympathetic ganglia and ganglionic cells

Muscularis

FIGURE 10.30

LM of parasympathetic ganglionic cell bodies in the muscularis layer of the gastrointestinal tract. The cell bodies of the parasympathetic ganglia are generally located near or within the effector organs. The cells may form either a small cluster of neurons in the organ or a well-organized structure as seen in the otic ganglia (spiral ganglia). Like other ganglia, the cell bodies have large nuclei and basophilic cytoplasm. The neurons are surrounded by afferent and efferent fibers, a few connective tissue fibers, blood vessels, and support cells. (400×)

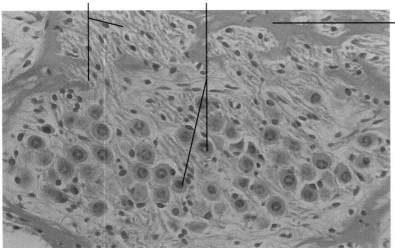

Cochlear nerve fibers

Bipolar cell bodies

Osseous cochlea

FIGURE 10.31
LM of spiral ganglia as seen in the spongy bone modiolus located in the cochlea of the inner ear. The spiral ganglia are composed of bipolar afferent neurons. The long axons of the bipolar neurons converge together and form the cochlea nerve. Short dendrites from the neurons articulate with the hair cells located in the hearing apparatus, the organ of **Corti.** (200×)

FIGURE 10.32
LM of choroid plexus derived from the highly vascularized region of the pia mater called the tela choroidea, and the ependymal cells lining the ventricles of the brain. The ependymal cells of the choroid plexus are cuboidal in shape, and may show numerous microvilli and cilia on their laminar surfaces. (100×)

Ventricle

Choroid plexus

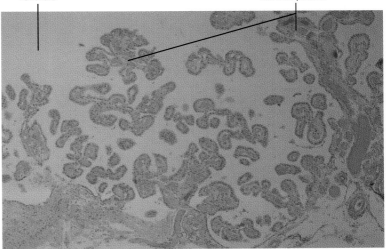

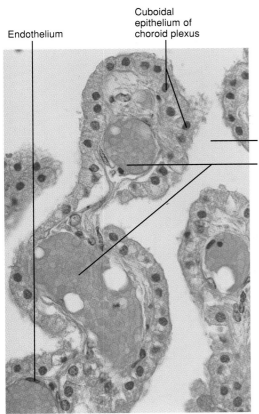

Endothelium

Cuboidal epithelium of choroid plexus

Ventricle

Wide fenestrated capillaries

FIGURE 10.33
LM of choroid plexus at a higher magnification. Note the microvilli lining the cuboidal epithelium, and protruding into the lumen of the ventricle. Beneath the epithelial lining are numerous thin-walled blood capillaries containing large numbers of erythrocytes. The choroid plexus secretes cerebrospinal fluid. (200×)

Central Nervous System

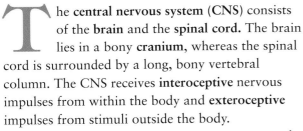

The **central nervous system (CNS)** consists of the **brain** and the **spinal cord.** The brain lies in a bony **cranium,** whereas the spinal cord is surrounded by a long, bony vertebral column. The CNS receives **interoceptive** nervous impulses from within the body and **exteroceptive** impulses from stimuli outside the body.

The cells of the nervous tissue—neurons and neuroglia cells—have been described in the previous chapter. This chapter will cover **membranes** of the CNS, **transverse** sections of the spinal cord through the **cervical, thoracic, lumbar,** and **sacral** regions, and the morphology associated with the **cerebrum** and **cerebellum.**

Membranes of the CNS Protection for the CNS is provided by a bony case, the cranium, which covers the brain, a vertebral column that covers the spinal cord, and a three-membrane investment called the meninges that covers both the brain and the spinal cord. The outermost investment of the meninges, which blends with the bony structure lying above it, is a tough, fibrous, rela-

tively inelastic membrane called **dura mater** or **pachymeninx.** The middle membrane of the meninges is the **arachnoid,** composed of a reticulate fiber network, **trabeculae,** and a subarachnoid space. The innermost membrane of the meninges is the **pia mater,** a thin membrane that closely invests the brain and extends into the depths of the **cerebral sulci.** The pia mater also surrounds the entire spinal cord and extends into the depths of the **anterior median fissure.**

Spinal cord In cross section, the spinal cord displays an **H-shaped** central area of **gray matter** composed of nerve cells and their fibers, and a central canal lined by **ependymal cells.**

The **white matter** of the spinal cord is composed of nerve fibers surrounding the gray matter. At different levels of the spinal cord—in the **cervical, thoracic, lumbar,** and **sacral** regions—there are variations in the shapes of the white and the gray matter. However, the basic structure of the cord is similar: gray matter surrounded by white matter.

The length of the cord is traversed anteriorly (ventrally) by a deep longitudinal cleft, the **anterior median fissure,** and posteriorly by the **dorsal median septum.** The pia mater invests the entire cord and extends deep into the anterior median fissure.

Cerebrum   Gray matter forms the upper surface of the cerebrum (**cerebral cortex**). The cortical surface forms convolutions called **gyri** that increase the surface area of the gray matter. Between the convoluted folds, or gyri, are depressions called **sulci.** The cortex contains nerve and neuroglia cells, fibers, and blood vessels. The cortical cells are **stellate, pyramidal, fusiform,** or **spindle-shaped,** and are generally arranged in stratified layers.

In a longitudinal section of the cortex, six distinct layers can be identified. Beginning at the outer surface, these are: the **molecular layer** (mostly fibrous with few nerve bodies), the **external granular** layer, the **pyramidal cell** layer, the internal **granular** layer, the internal **pyramidal** or **ganglionic** layer, and the **multiform** or **polymorphic cell** layer.

Underlying the white matter are bundles of **myelinated and nonmyelinated neuronal fibers,** which extend in all directions. The fibers form three main groups: the **association** fibers, the **commis-**sural fibers, and the **projection** fibers. Between the fibers are **neuroglia** cells.

Cerebellum   The cerebellum coordinates movements of striated muscle and maintains equilibrium and posture. The cerebellum is divided into left and right **hemispheres.** Between the hemispheres is the **vermis,** which is segmented into lobules separated by transverse fissures. As in the case of the cerebrum, the gray matter in the cerebellum is located on the surface, forming a thin layer of **cerebellar cortex.** Below the cortex lies the white matter with myelinated fibers, and a few nerve cells lie in the middle of the cerebellum.

In a sagittal section the cerebellar cortex displays three layers: an outer **molecular** layer in which few small nerve cells and many nonmyelinated fibers can be identified, a middle layer with large **Purkinje** (neuron) cells with fibrous dendrites, and the inner **granular** layer of small nerve cell bodies. The nerve cells generally have three to six dendrites associated with the cell body, and a **nonmyelinated** axon that traverses toward the molecular layer and then splits into two lateral branches that run along the length of the **folium.**

Dorsal Aspect of the Brain

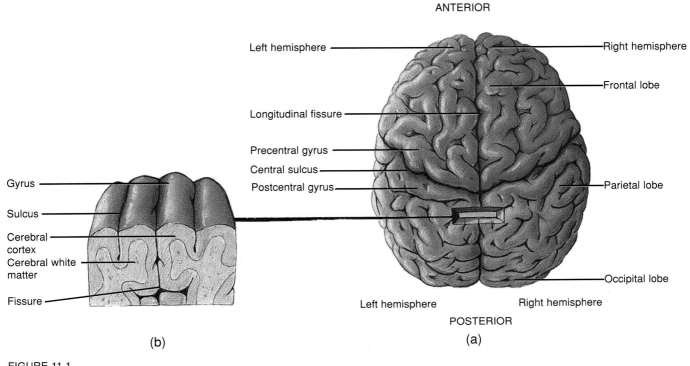

FIGURE 11.1
(a) Diagram of dorsal aspect of the human brain, illustrating major regions and structures.
(b) Diagram of an inset from the occipital lobes, demonstrating the differences among the gyrus, sulcus, cerebral cortex, cerebral white matter, and fissure.

Ventral Aspect of the Brain and the Brainstem

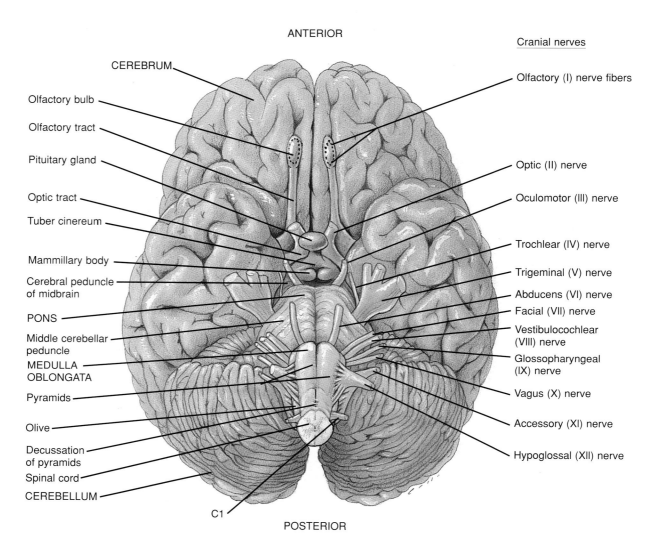

ANTERIOR

CEREBRUM

Olfactory bulb

Olfactory tract

Pituitary gland

Optic tract

Tuber cinereum

Mammillary body

Cerebral peduncle
of midbrain

PONS

Middle cerebellar
peduncle

MEDULLA
OBLONGATA

Pyramids

Olive

Decussation
of pyramids

Spinal cord

CEREBELLUM

C1

POSTERIOR

Cranial nerves

Olfactory (I) nerve fibers

Optic (II) nerve

Oculomotor (III) nerve

Trochlear (IV) nerve

Trigeminal (V) nerve

Abducens (VI) nerve

Facial (VII) nerve

Vestibulocochlear
(VIII) nerve

Glossopharyngeal
(IX) nerve

Vagus (X) nerve

Accessory (XI) nerve

Hypoglossal (XII) nerve

FIGURE 11.2
Diagram of ventral aspect of the human brain and brainstem, illustrating major regions and structures.

Cerebral Cortex

FIGURE 11.3
Light micrograph (LM) of a sagittal section through a portion of a cerebral hemisphere. The gray matter of the cerebrum is located on the surface of the cerebral cortex. The white matter lies below the cortex as bundles of myelinated and nonmyelinated fibers, spreading in all directions, and neuroglia cells. In general, the surfaces of the hemispheres are highly convoluted, which increases the relative surface area of the hemispheres. The projection folds of the hemispheres are the gyri, and the depressions are the sulci. The cerebral cortex contains nerve cells, neuroglia, fibers, and blood vessels. The cerebral hemisphere is invested by pia mater. Silver stain. (20×)

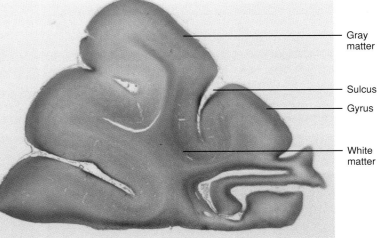

Gray matter

Sulcus

Gyrus

White matter

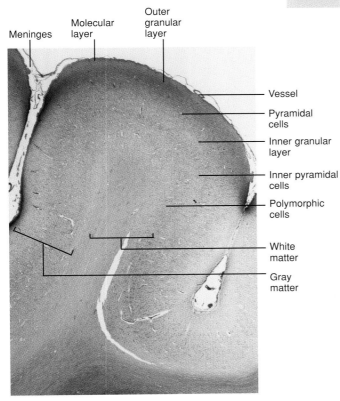

Meninges

Molecular layer

Outer granular layer

Vessel

Pyramidal cells

Inner granular layer

Inner pyramidal cells

Polymorphic cells

White matter

Gray matter

FIGURE 11.4
LM of a sagittal section through a gyrus. Covering the gyrus is a thin veneer of meninges. Lateral to the gyrus are invaginations of the sulci. Different shades of brown and gray represent the stratification of nerve cells and their fibers. The white streak in the central part of the tissue is an artifact. Silver stain. (20×)

FIGURE 11.5
LM of a sagittal section through the cerebral cortex. Visible are six layers of cells and fibers from the superficial layer to the deep layer of the cortex. These are: the molecular layer, composed of fibers originating from cells in the lower layers, and few neurons; the external granular layer, composed of small, triangular nerve cell bodies; the external pyramidal layer, with large pyramidal cells mixed with small granule cells; the internal granular layer, composed of small stellate granule cells; the internal pyramidal or ganglionic layer, with large and medium-sized pyramidal cells; and the multiform or polymorphic cell layer, composed of cells of diverse shapes and sizes. (100×)

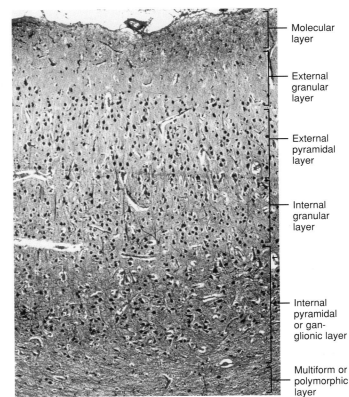

Molecular layer

External granular layer

External pyramidal layer

Internal granular layer

Internal pyramidal or ganglionic layer

Multiform or polymorphic layer

FIGURE 11.6

LM of a sagittal section through the pyramidal layer of the cerebral cortex. The pyramidal cells direct their dendrites toward the surface of the cortex and are therefore called apical dendrites. The larger pyramidal cells (the largest of their kind in the cortex) are located in the motor cortex and are called **Betz** cells. The area between the glial and pyramidal cells is the neuropil area, which is filled by neuroglial cell processes, dendrites, and axonal processes of nerve cells. (200×)

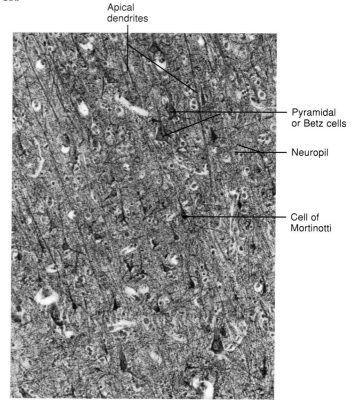

Apical dendrites

Pyramidal or Betz cells

Neuropil

Cell of Mortinotti

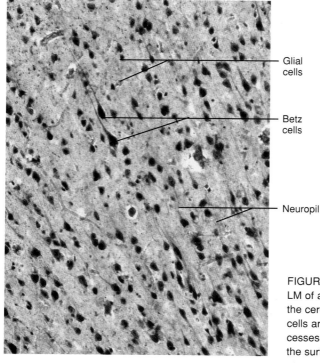

Glial cells

Betz cells

Neuropil

FIGURE 11.7

LM of a sagittal section, at a higher magnification, through the pyramidal layer of the cerebral cortex. The micrograph shows giant pyramidal **Betz** cells. The Betz cells are surrounded by neuropil, a feltwork of nonmyelinated fibers and processes of neuroglia cells. The apical dendrites of pyramidal cells extend toward the surface of the cortex. (400×)

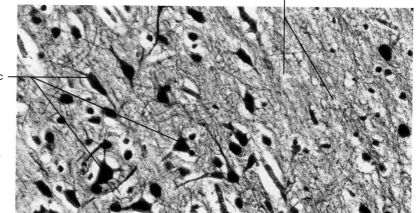

Neuropil

Polymorphic cells

FIGURE 11.8

LM of a sagittal section through the multiform (polymorphic) layer of the cerebral cortex. Cells of the multiform layer vary in shape and size. The cell bodies of neurons in this layer are larger than those of neurons in more superficial layers. Neuropil surrounds the cell bodies. (400×)

Cerebellum

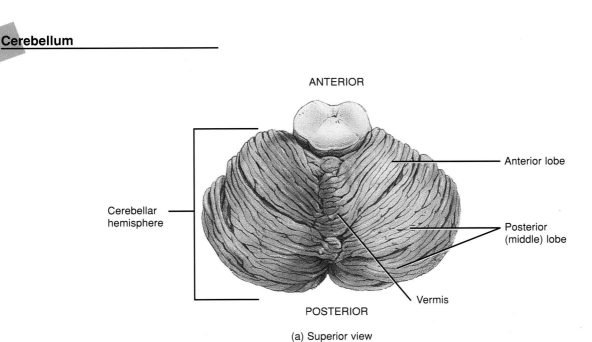

ANTERIOR

Cerebellar
hemisphere

Anterior lobe

Posterior
(middle) lobe

POSTERIOR

Vermis

(a) Superior view

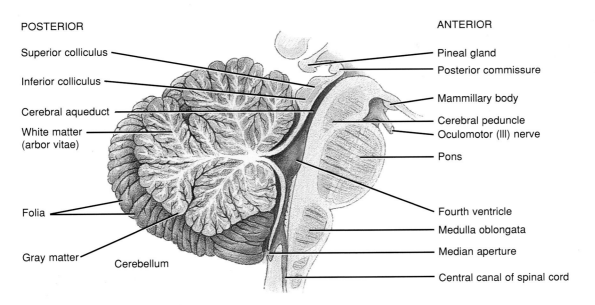

POSTERIOR

ANTERIOR

Superior colliculus

Inferior colliculus

Cerebral aqueduct

White matter
(arbor vitae)

Folia

Gray matter

Cerebellum

Pineal gland
Posterior commissure

Mammillary body

Cerebral peduncle
Oculomotor (III) nerve

Pons

Fourth ventricle

Medulla oblongata

Median aperture

Central canal of spinal cord

(b) Sagittal section of cerebellum and brain stem

FIGURE 11.9
Cerebellum.

Fissures

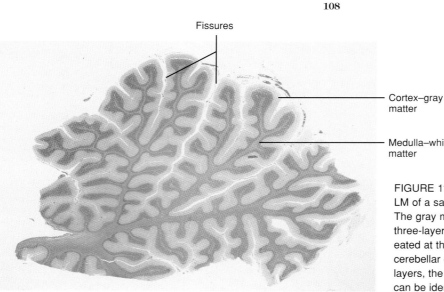

Cortex–gray matter

Medulla–white matter

FIGURE 11.10
LM of a sagittal section through the cerebellar cortex. The gray matter of the cortex has a simple, uniform, three-layer structure. The three layers cannot be delineated at this magnification. In the section, the outer cerebellar cortex, the inner white matter (arbor vitae) layers, the primary fissures, and the semilunar lobules can be identified. Pia mater covers the cerebellum. (1×)

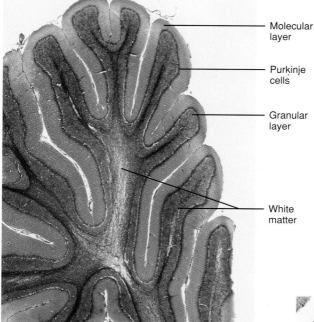

Molecular layer

Purkinje cells

Granular layer

White matter

FIGURE 11.11
LM of a sagittal section through the cerebellar cortex. The micrograph shows the outer molecular layer, which contains a few small neurons and nonmyelinated fibers. Below this is the **Purkinje** cell layer of large, flask-shaped neurons called Purkinje cells. The inner granular layer of the gray matter consists of abundant small neurons. White matter (arbor vitae) lies below the gray matter. (20×)

Molecular layer

Purkinje cells

Granular layer

White matter

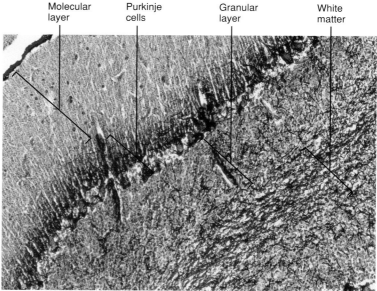

FIGURE 11.12
LM of a sagittal section through the cerebellar cortex. The micrograph shows the superficial molecular layer, which consists of dendrites and nonmyelinated axons of various cerebellar **Purkinje** neurons, stellate cells, and basket cells. The layer below the molecular layer is a single row of large, flask-shaped neurons or Purkinje cells. Dendrites from these neurons arborize profusely in the molecular layer, whereas the axons enter the medullary layer. The granular layer lies adjacent to the medullary area and is composed of stratified granule cells (small neurons). The granule cells receive major cerebellar input from the mossy fibers of the Purkinje cells. Silver stain. (200×)

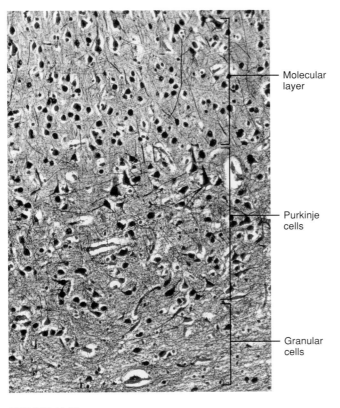

Molecular layer

Purkinje cells

Granular cells

FIGURE 11.13
LM of a small area at the junction of the molecular and **Purkinje** cell layers. Stellate (basket) cells, granule cells, glomeruli (islands) between cells, and neuronal fibers can be identified in this section. Silver stain. (200×)

Molecular layer

Purkinje cell bodies

Granular layer

Granule cells

FIGURE 11.14
LM of a sagittal section through the molecular and **Purkinje** cell layers of the cerebellar cortex. Purkinje cells can be seen between the granular and molecular layers. Cell bodies of the Purkinje cells have distinct nuclei, which are difficult to see in silver-impregnated tissue. **Nissl** stain. (100×)

FIGURE 11.15
LM of **Purkinje** cells stained by a silver-impregnation technique. The Purkinje cells form a single row at the junction of the granular and molecular layers. Branching extensively, fan-shaped dendrites of the Purkinje cells reach up to the cerebellar surface. The axon extends from the large, flask-shaped cell body and enters the white matter, where it is myelinated by neuroglia cells. The cell body is enmeshed in a delicate network of nerve fibers of various origins, forming a basket-like appearance around the cell body, which gives rise to the term "basket cell." (200×)

Spinal Cord

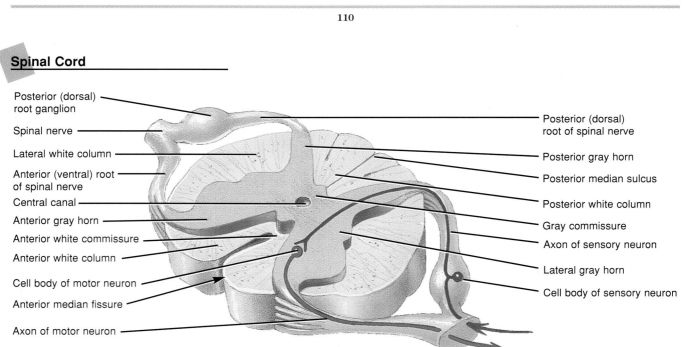

Posterior (dorsal) root ganglion

Spinal nerve

Lateral white column

Anterior (ventral) root of spinal nerve

Central canal

Anterior gray horn

Anterior white commissure

Anterior white column

Cell body of motor neuron

Anterior median fissure

Axon of motor neuron

Posterior (dorsal) root of spinal nerve

Posterior gray horn

Posterior median sulcus

Posterior white column

Gray commissure

Axon of sensory neuron

Lateral gray horn

Cell body of sensory neuron

FIGURE 11.16
Diagram of the organization of gray and white matter in the spinal cord, as seen in cross section. Note the microscopic components of the posterior root ganglion, the posterior root of the spinal nerve, the anterior root of the spinal nerve, and the spinal nerve.

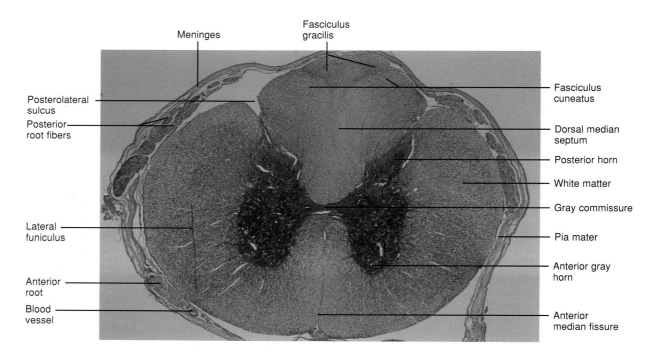

Meninges

Fasciculus gracilis

Posterolateral sulcus

Posterior root fibers

Lateral funiculus

Anterior root

Blood vessel

Fasciculus cuneatus

Dorsal median septum

Posterior horn

White matter

Gray commissure

Pia mater

Anterior gray horn

Anterior median fissure

FIGURE 11.17
LM of a cross section of the spinal cord in the cervical region, illustrating the organizational arrangement of gray and white matter and the surrounding meninges. The meninges consists of an outer dura mater, a middle arachnoid, and an inner pia mater. The inner dark brown butterfly-shaped gray matter contains cell bodies of sensory, motor, and association neurons and their fibers. In the center of the section is the central canal of the spinal cord lined by ependymal epithelial cells. The gray matter also displays a connecting bridge of gray commissure fibers. The gray matter extends anteriorly, forming prominent anterior horns where large cell bodies of motor neurons are located. Smaller, sensory neurons are present in the posterior horns. The two invaginations, the anterior median fissure and the posterior median sulcus, partially divide the spinal cord into two halves. The white matter of the spinal cord is composed of myelinated and nonmyelinated nerve fibers forming longitudinal columns, or funiculi, and neuroglia cells. The white matter surrounds the gray matter in the micrograph. (20×)

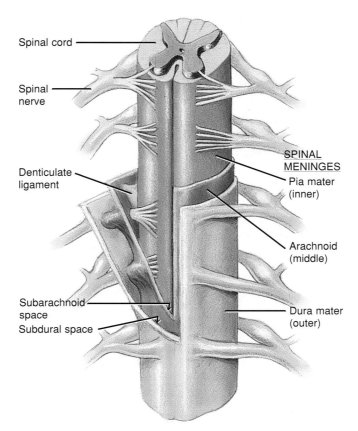

Spinal cord

Spinal nerve

Denticulate ligament

SPINAL MENINGES
Pia mater (inner)

Arachnoid (middle)

Subarachnoid space
Subdural space

Dura mater (outer)

FIGURE 11.18
Diagram of a small region of the spinal cord, showing the spinal cord covering and the subdivisions of the spinal meninges—the pia mater, the arachnoid, and the dura mater. Also shown are the spinal nerve, the denticulate ligament, and the subarachnoid and subdural spaces.

Upper Cervical or Caudal Region of the Medulla

FIGURE 11.19
LM of a cross section through the spinal cord at the upper cervical or caudal region of the medulla. The micrograph identifies the spinal trigeminal tract area and the spinal trigeminal nucleus. The dorsal median sulcus and the dorsal intermediate sulcus can also be identified in the section. (1×)

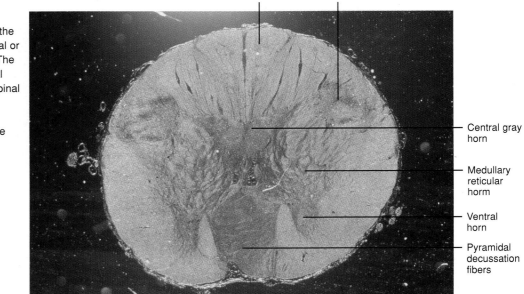

Fasciculus gracilis and nucleus

Trigeminal spinal nucleus and tract

Central gray horn

Medullary reticular horn

Ventral horn

Pyramidal decussation fibers

Lower Cervical

FIGURE 11.20

LM of a cross section through the lower cervical spinal cord in the region of the eighth cevical vertebra (C8). The white matter surrounds the gray matter, a characteristic of the spinal cord. The dorsal median septum, ventral median fissure, posterior funiculus, posterior horn, lateral funiculus, intermediate gray horn, anterior horn, and anterior funiculus can be identified in the micrograph. (1×)

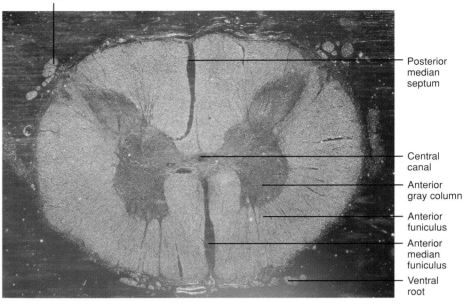

Dorsal root

Posterior median septum

Central canal

Anterior gray column

Anterior funiculus

Anterior median funiculus

Ventral root

Thoracic

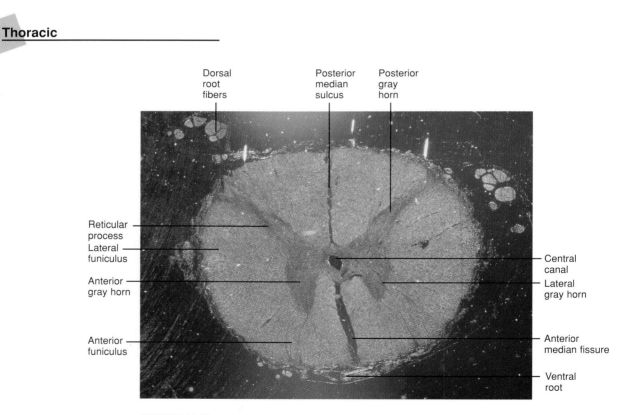

Dorsal root fibers

Posterior median sulcus

Posterior gray horn

Reticular process

Lateral funiculus

Anterior gray horn

Anterior funiculus

Central canal

Lateral gray horn

Anterior median fissure

Ventral root

FIGURE 11.21

LM of a cross section through the thoracic region of the spinal cord. The characteristic H-shaped arrangement of the gray matter surrounded by the white matter can be seen in the micrograph. Other structures that can be identified are the reticular process, the anterior, lateral, and posterior gray horns, the anterior funiculus, the central canal, the posterior median sulcus, the anterior median fissure, and the dorsal root fibers. A distinct nuclear mass, the nucleus dorsalis, is visible at the base of the posterior horn. (1×)

Lumbar

FIGURE 11.22
LM of a cross section of the spinal cord in the lumbar region. The H-shaped gray matter is surrounded by the white matter. Massive concentrations of gray matter are characteristic of this region of the spinal cord. Seen in the micrograph are the funiculi, dorsal, and ventral roots, substantia gelatinosa, posterior median septum, and anterior median fissure of the spinal cord. (1×)

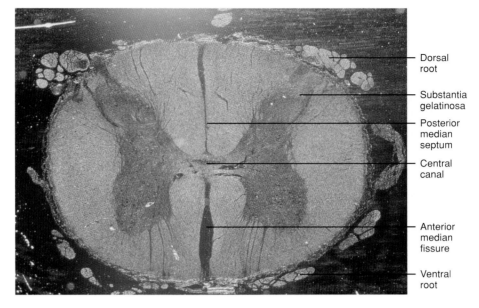

Dorsal root

Substantia gelatinosa

Posterior median septum

Central canal

Anterior median fissure

Ventral root

Fasciculus gracilis

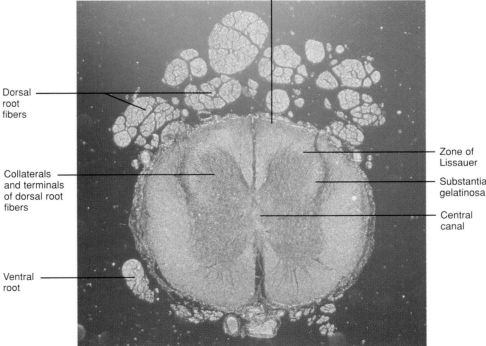

Dorsal root fibers

Collaterals and terminals of dorsal root fibers

Ventral root

Zone of Lissauer

Substantia gelatinosa

Central canal

FIGURE 11.23
LM of a cross section of the spinal cord through the sacral region. Bundles of myelinated nerve fibers originating in the dorsal root ganglia enter the cord. Also visible in the micrograph are the dorsal root fibers, myelinated fibers of the fasciculus dorsolateralis, dorsal and ventral root nerve fibers, central canal, ventral white commissure, fasciculus gracilis, posterior median septum, and anterior medium fissure. (1×)

Blood
vessels   Nerves

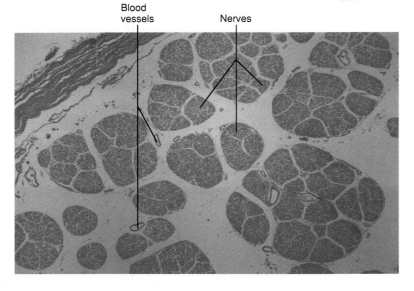

FIGURE 11.24
LM of a cross section of the cauda equina (Latin for
"horse's tail"). The cauda equina is a collection of
dorsal and ventral roots (nerves) extending from the L1
and L2 vertebrae to the end of the S2 region of the
sacrum. The cross section shows nerves, blood vessels, and the surrounding meninges. (100×)

FIGURE 11.25
LM of a cross section of a nerve in the cauda
equina region of the spinal cord. Note myelinated fibers, axons, and neuroglia cells in the
region. (400×)

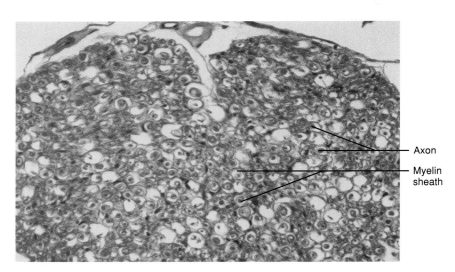

Axon

Myelin
sheath

Neuroglia   Nucleus

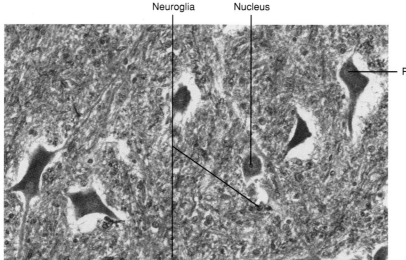

Perikaryon

FIGURE 11.26
LM of a cross section through the gray matter of the
spinal cord in the lower cervical region. The micrograph shows large perikaryons of multipolar neurons.
Surrounding the neurons are neuroglia cells with their
fibers, and fibers originating from the neurons in the
gray matter. (400×)

CHAPTER 12

Integumentary System

The integumentary system (skin and related structures) has the greatest mass of all the systems in the body. It functions as a barrier of protection from mechanical, chemical, and temperature-related injuries. The integument also functions in fluid homeostasis, temperature regulation, excretion of wastes, and neurosensory input to effector organs of the body. The integument consists of two components: the **skin,** which includes the **epidermis** and **dermis,** and the underlying **hypodermis.**

The epidermis is an epithelium several cell layers thick. It has an underlying noncellular basement membrane. The epidermis consists of squamous cells of various shapes. There are also several specialized cells within the epidermis.

Five different types of cells form different layers in the epidermis. These layers include the **stratum basale,** the **stratum spinosum,** the **stratum granulosum,** the **stratum lucidum** (only in thick skin), and the **stratum corneum.**

The **stratum basale** (also called **stratum germinativum**) lies superior to the dermis and is con-

tiguous with the basement membrane of the epidermis. The stratum basale consists of a single row of columnar cells. Mitotic figures are found within the basale layer. **Melanocytes,** pigment-secreting cells, are also present in this layer.

The **stratum spinosum** is the next microanatomical layer of the epidermis. It is superficial to the basale layer. The stratum spinosum consists of several layers of **polyhedral** shaped cells. Present are **desmosomes,** which attach several cells together. These structures pull on the cell membranes and produce a prickly appearance in the cell layers.

The **stratum granulosum** lies above the stratum spinosum. The stratum granulosum is approximately five layers thick. The cells within this layer contain distinct **keratohyalin granules,** which become more prominent as the cells approach the surface. The nuclei in the cells become less distinct as these cells approach the overlying stratum lucidum.

The **stratum lucidum** is superficial to the stratum granulosum. The squamous cells are flat and lack nuclei and other cellular organelles. The lucidum layer is only found in thick skin of the palm and foot. The **stratum corneum** is several layers thick. The cells in this layer lack organelles, nuclei, and cytoplasm.

The **dermis** lies below the epidermis and is composed of **collagenous** and **elastic connective tissue:** fibroblasts, mast cells, macrophage, blood vessels, and nerve endings. The dermis consists of two microscopically different layers: the **reticular** and **papillary** layers. The **reticular** layer lies below the papillary layer and is composed of more densely arranged connective tissue. Cell types are less abundant in this region than in the papillary region. The **papillary** region of the dermis consists of less densely arranged connective tissue and a wider variety of cell types. The papillary layer forms invaginations extending into the overlying epidermis. The invaginations are called **dermal papillae.**

The **hypodermis,** or superficial facia, lies below the dermis. It is composed of **loose areolar connective tissue** surrounding adipose tissue. It adheres to the dermis by means of collagenous fibers.

The skin appendages are modifications of the epidermis and consist of **nails, sweat glands (eccrine** and **apocrine), sebaceous glands,** and **hair.**

The **nails** are keratinized plaques and consist of several layers of **cornified epithelium.** The nail consists of the **nail bed,** the **nail root,** the **nail plate,** the **nail fold,** the **eponychium,** and the **hyponychium.**

Sweat glands are of two types: eccrine and apocrine. The **eccrine glands** are simple tubular structures consisting of **cuboidal** to low **columnar epithelium.** The epithelium is adherent to a relatively thick basement membrane. The main secretory portion of the gland is a coiled structure located in the dermis. The secretory portion of the gland empties into the eccrine duct, which is lined with two layers of cuboidal epithelium. The secretory epithelium of the gland is cuboidal with two cell types: **dark cells** and **clear cells.** In addition, there are special **myoepithelial** cells located around the secretory structures.

Apocrine glands are somewhat different. They are usually larger and the epithelium has only one cell type in its secretory parenchyma. The apocrine gland empties into a **hair follicle.** Thus, the ducts of these glands are composed of **stratified squamous epithelium.**

Sebaceous glands are more complex than the simple tubular sweat glands. The **parenchymal epithelium** consists predominantly of **polyhedral** cells arranged in the form of **alveoli.** The sebaceous alveoli are located in the dermis and are surrounded by thin connective tissue. The alveoli of the sebaceous glands usually drain their secretions into the superior portions of hair follicles. Therefore, the sebaceous ducts are also lined by **stratified squamous epithelium.**

Hair follicles are relatively complex skin appendages with three basic layers: a **connective tissue sheath,** and **outer** and **inner epithelial root sheaths.** The hair follicle connective tissue sheath is composed of three layers: an **inner glassy membrane,** a **middle connective tissue layer,** and an outer layer of mixed collagenous and elastic fibers.

The **outer** and **inner epithelial root sheaths** are continuous with the epidermis. Morphologically, the outer epithelial root sheath is a continuation of the **stratum basale** and the **stratum spinosum** (collectively, the **Malpighii layer).** The inner layer of the root sheath is a modification of the superficial layer of the epidermis. It is composed of the **cuticle** of the **root, Huxley's layer,** and Henle's layer.

The hair shaft is composed of **keratinized epithelial cells.** It has both a **medulla** and a **cortex.** The medulla consists of predominantly **polyhedral** cells. The cortex consists of elongated cells.

The **integument** contains specialized sensory structures. These structures are the **Pacinian corpuscle,** the **Meissner's corpuscle,** and **free nerve endings.** Pacinian corpuscles are onion-like laminar structures located in the **reticular dermis. Meissner's corpuscles** are ovoid structures located in the **papillary dermis. Free nerve endings** may be found in the epidermis as well as in the dermis. In the dermis, they may radiate fibers to the **hair papillae.**

The Skin and Related Structures

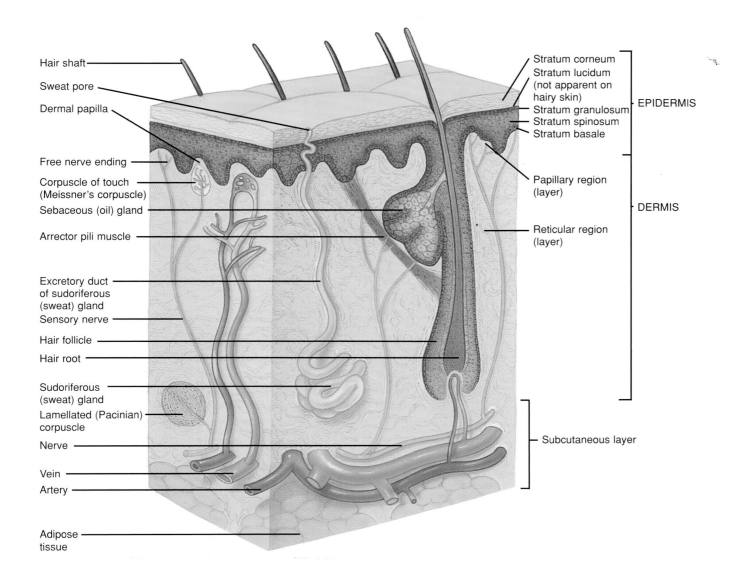

Hair shaft
Sweat pore
Dermal papilla

Free nerve ending
Corpuscle of touch
(Meissner's corpuscle)
Sebaceous (oil) gland
Arrector pili muscle

Excretory duct
of sudoriferous
(sweat) gland
Sensory nerve
Hair follicle
Hair root
Sudoriferous
(sweat) gland
Lamellated (Pacinian)
corpuscle
Nerve
Vein
Artery

Adipose
tissue

Stratum corneum
Stratum lucidum
(not apparent on
hairy skin)
Stratum granulosum
Stratum spinosum
Stratum basale

EPIDERMIS

Papillary region
(layer)

DERMIS

Reticular region
(layer)

Subcutaneous layer

FIGURE 12.1
Structure of the skin and underlying subcutaneous tissue.

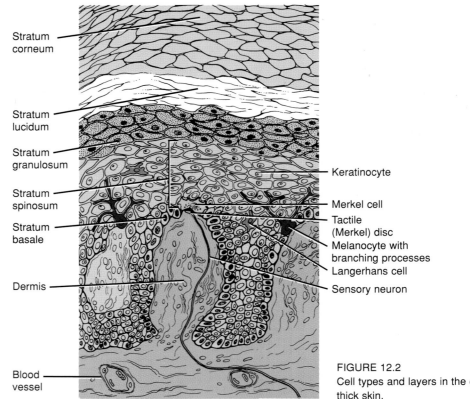

Stratum corneum

Stratum lucidum

Stratum granulosum

Stratum spinosum

Stratum basale

Dermis

Keratinocyte

Merkel cell

Tactile (Merkel) disc

Melanocyte with branching processes

Langerhans cell

Sensory neuron

Blood vessel

FIGURE 12.2
Cell types and layers in the epidermis of thick skin.

Blood vessel

FIGURE 12.3
Light micrograph (LM) of a section through thick skin as seen in the palm of the hand or the soles of the feet. The skin in these regions is heavily cornified. This section shows the epidermis, the dermis, and the hypodermis, as well as several arterioles, a **Pacinian** corpuscle, and a glomus. (40X)

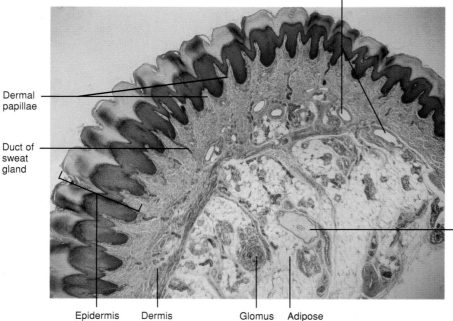

Dermal papillae

Duct of sweat gland

Pacinian corpuscle

Epidermis   Dermis   Glomus   Adipose tissue

Stratum
granulosum

Stratum
corneum

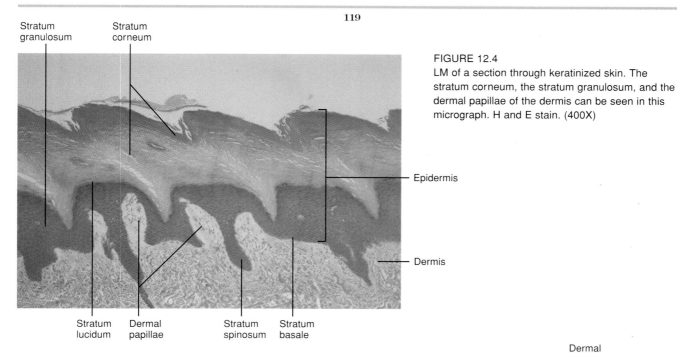

Epidermis

Dermis

Stratum
lucidum

Dermal
papillae

Stratum
spinosum

Stratum
basale

FIGURE 12.4
LM of a section through keratinized skin. The stratum corneum, the stratum granulosum, and the dermal papillae of the dermis can be seen in this micrograph. H and E stain. (400X)

FIGURE 12.5
LM of a section of skin, showing the epidermis, the dermis, and the hypodermis. Rete pegs are prominent. Dermal papillae, sweat glands, and adipose tissue in the hypodermis can be identified in this section. (100X)

Dermal
papillae

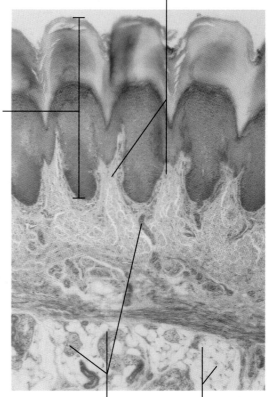

Epidermis

Sweat
glands

Adipose
tissue

Dermis

Stratum
spinosum

Stratum
granulosum

Stratum
corneum

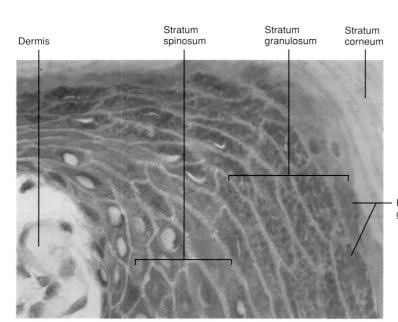

Basophilic
granules

FIGURE 12.6
LM of skin at a higher magnification. Stratum corneum, stratum granulosum, and stratum spinosum can be seen in the micrograph. Also, note the heavy granular appearance of the stratum granulosum. H and E stain. (1000X)

FIGURE 12.7
LM of a section through skin, showing the
stratum spinosum at a higher magnification. Note the intercellular, delicate cytoplasmic bridges, which give the cells a
prickly appearance. Nuclei of spinosum
cells are prominent in the micrograph. H
and E stain. (1000X)

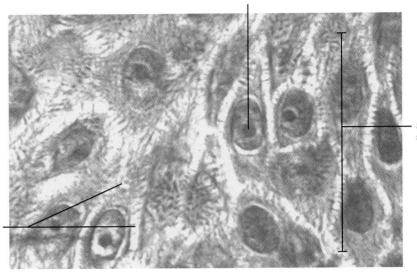

Nucleus

Stratum
spinosum

Cytoplasmic
spines

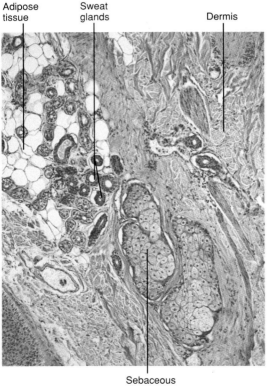

Adipose
tissue

Sweat
glands

Dermis

Sebaceous
gland

FIGURE 12.8
LM of a section through the dermis of the skin. This
section shows numerous sweat glands and ducts
near the hypodermis. Sebaceous glands are
located more centrally in the micrograph. Note the
fine connective tissue capsule surrounding the
sebaceous glands. H and E stain. (100X)

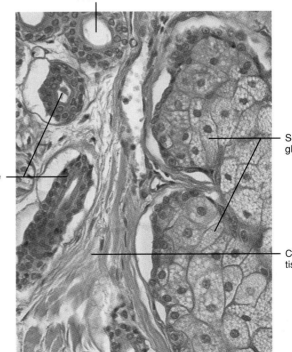

Excretory
duct

Sebaceous
glands

Merocrine
sweat
glands

Connective
tissue

FIGURE 12.9
LM of skin section at a higher magnification. In the micrograph,
sebaceous glands appear as foamy structures. The cells are polyhedral with central nuclei. Merocrine sweat glands with cuboidal
cells secrete a watery fluid (sweat) that is rich in sodium and chlo-
ride ions, small amounts of urea, and other metabolites. A fine connective tissue surrounds the gland. Sweat glands are to the left in
this micrograph. H and E stain. (400X)

FIGURE 12.10
LM of a section through skin in the axilla region. This section shows two distinct types of sweat glands: merocrine sweat glands and apocrine sweat glands. Apocrine sweat glands are found mainly in the axillae and the genital region. These glands are much larger in size and have larger lumina compared with the merocrine sweat glands. Also seen in the micrograph are adipose and connective tissue. (100X)

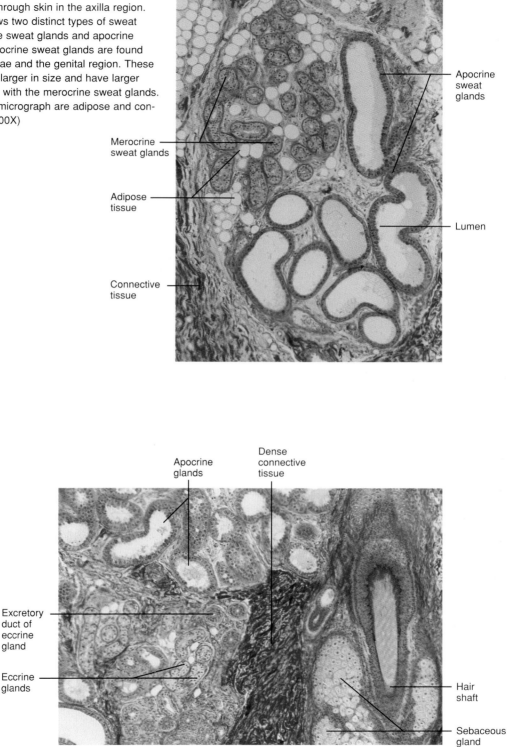

FIGURE 12.11
LM of a section through skin in the axilla region. This section shows sebaceous glands near a hair shaft and apocrine and merocrine glands to the left. Smaller eccrine glands are noted below the apocrine glands. Also seen in the micrograph is dense, irregular connective tissue and structures associated with the hair shaft. H and E stain. (200X)

Hair and Hair Follicle

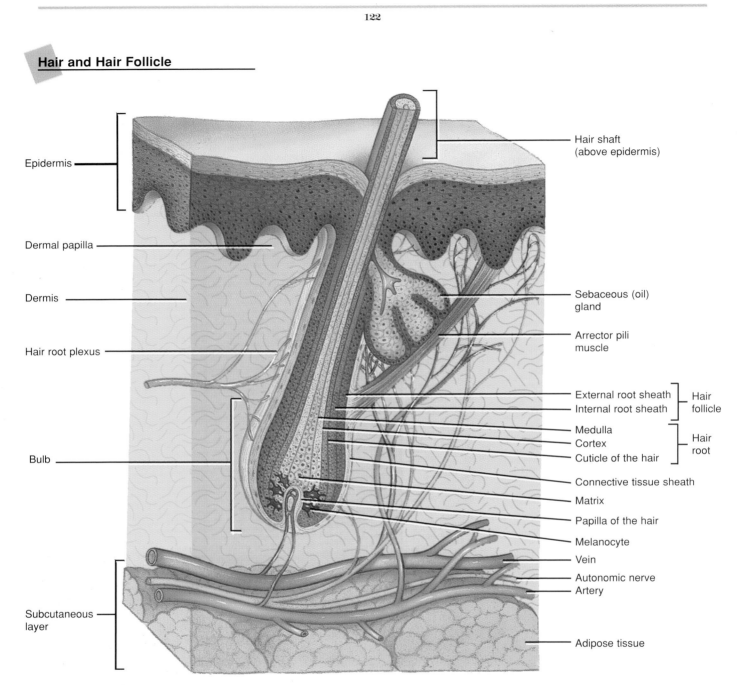

Epidermis

Dermal papilla

Dermis

Hair root plexus

Bulb

Subcutaneous layer

Hair shaft (above epidermis)

Sebaceous (oil) gland

Arrector pili muscle

External root sheath ⎤ Hair
Internal root sheath ⎦ follicle

Medulla ⎤
Cortex ⎬ Hair root
Cuticle of the hair ⎦

Connective tissue sheath

Matrix

Papilla of the hair

Melanocyte

Vein

Autonomic nerve

Artery

Adipose tissue

FIGURE 12.12
Diagrammatic cross-sectional representation of skin, showing the relationship of hair to the epidermis. Note that the hair follicle and the hair bulb lie deep in the dermis. Also shown are a sebaceous (oil) gland, an arrector pili muscle, a hair root, the subcutaneous layer, and supporting structures.

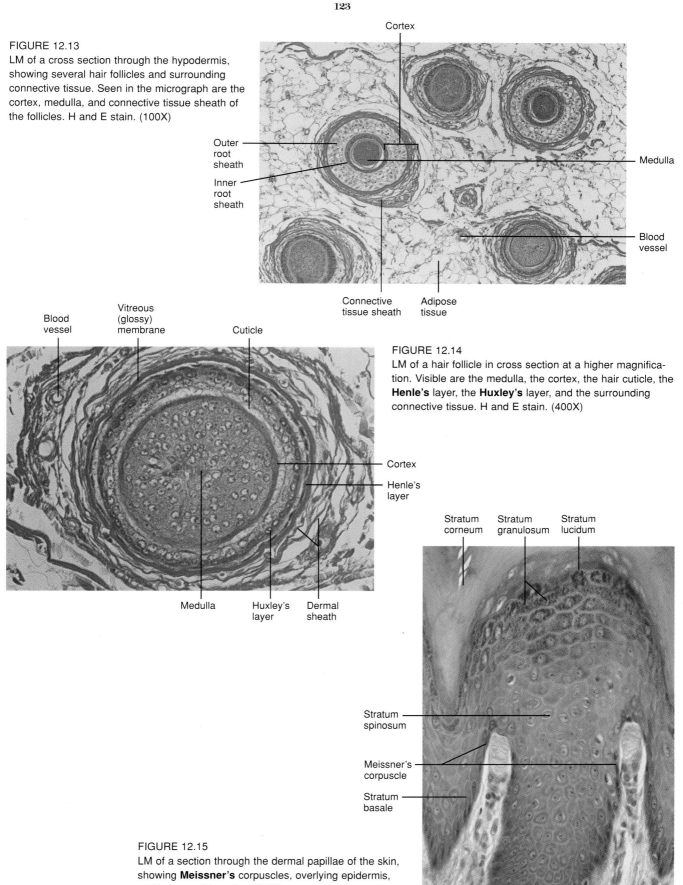

FIGURE 12.13
LM of a cross section through the hypodermis, showing several hair follicles and surrounding connective tissue. Seen in the micrograph are the cortex, medulla, and connective tissue sheath of the follicles. H and E stain. (100X)

FIGURE 12.14
LM of a hair follicle in cross section at a higher magnification. Visible are the medulla, the cortex, the hair cuticle, the **Henle's** layer, the **Huxley's** layer, and the surrounding connective tissue. H and E stain. (400X)

FIGURE 12.15
LM of a section through the dermal papillae of the skin, showing **Meissner's** corpuscles, overlying epidermis, and basement lamella. (400X)

Stratum
basale

Stratum
spinosum

Meissner's
corpuscle

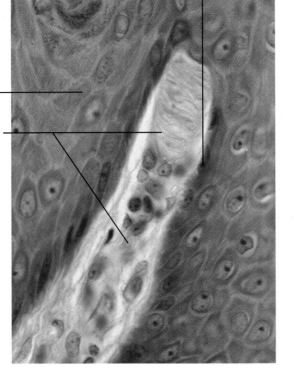

Excretory
duct

Dermis

Pacinian
corpuscles

Adipose
tissue

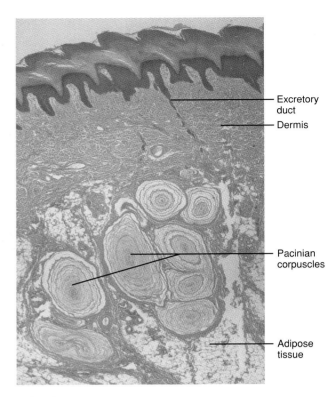

FIGURE 12.16
LM of a section through the dermis of the skin,
showing **Meissner's** corpuscles at a higher magnification. The structure is ovoid and has a layered disc
appearance. They lie in the dermal papillae and are
associated with nerve endings, and are involved in the
perception of discriminatory touch. Surrounding the
corpuscle is a fine network of connective tissue.
H and E stain. (1000X)

FIGURE 12.17
LM of a cross section through the skin. Visible are
the epidermis, the dermis, and the hypodermis.
There are several **Pacinian** corpuscles located in
the adipose tissue of the hypodermis. **Pacinian**
corpuscles are encapsulated sensory pressure
receptors for coarse touch, tension, and vibration.
They are located in the skin, ligaments, the serous
membranes, the joints, the mesenteries, the penis,
and some viscera. H and E stain. (40X)

FIGURE 12.18
LM of a cross section through the skin. Besides the regular cells in the connective tissue of the dermis, the
dermis in certain parts of the body may be highly pigmented because of branched pigmented cells called
chromatophores. This is especially true for dark-skinned
individuals. Chromatophores are also found in the
choroid coat of the eye, giving it a dark coloration.
(400X)

Branched
chromatophores

Dermis

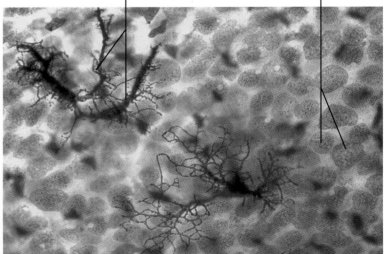

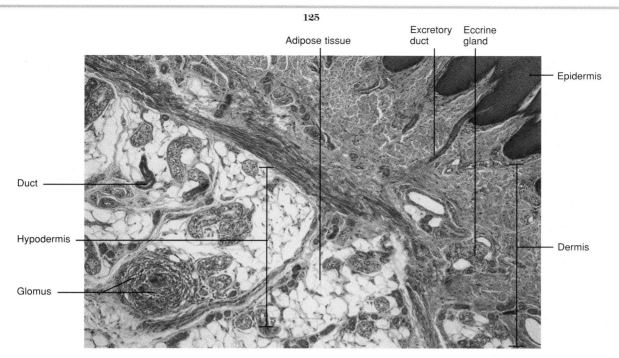

Adipose tissue

Excretory duct

Eccrine gland

Epidermis

Duct

Hypodermis

Glomus

Dermis

FIGURE 12.19
LM of a sagittal section of the skin of the fingertip. The micrograph shows the epidermis, the dermis, and the hypodermis. Adipose tissue, glands, and a glomus body (an arteriovenous shunt) can be seen in this section. (100X)

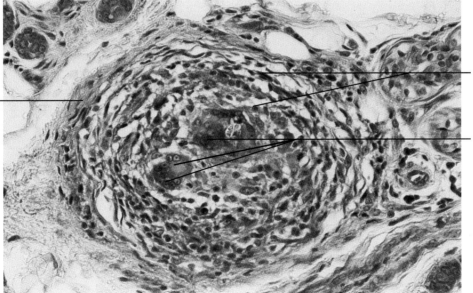

Capsule

Glomus media of epitheloid cells

Blood vessels

FIGURE 12.20
LM of a cross section through the skin. In the dermis of the fingertips and the external ear, there are specialized shunts called glomus bodies. The glomus is a highly convoluted vascular structure surrounded by a loose connective tissue capsule. Just prior to the arteriovenous junction, the arterial wall thickens and the smooth muscle cells in the body wall become epithelial in shape. (400X)

Structure of Nails

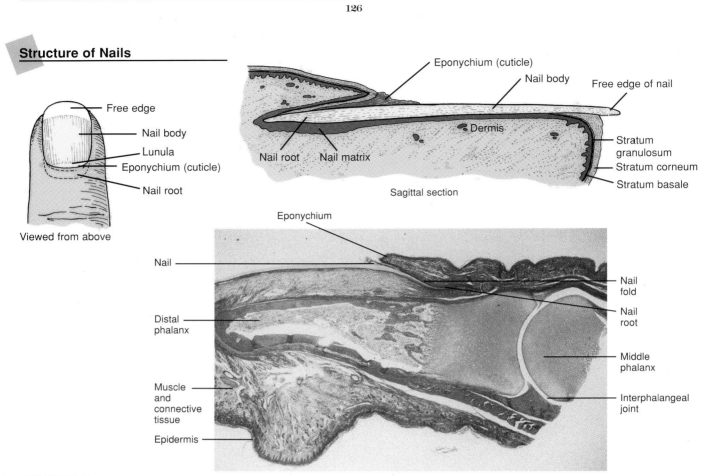

Free edge

Nail body

Lunula

Eponychium (cuticle)

Nail root

Viewed from above

Eponychium (cuticle)

Nail body

Free edge of nail

Dermis

Nail root

Nail matrix

Stratum granulosum

Stratum corneum

Stratum basale

Sagittal section

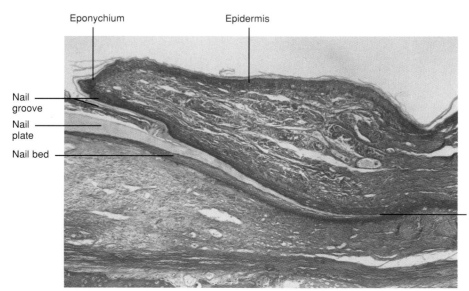

Eponychium

Nail

Distal phalanx

Muscle and connective tissue

Epidermis

Nail fold

Nail root

Middle phalanx

Interphalangeal joint

FIGURE 12.21

LM of a sagittal section through a fetal finger. On the anterior dorsal surface of the finger is a highly specialized appendage, the nail. The nail is a keratinized, dense plate that is attached to the stratified squamous epithelium called the nail bed. The nail root (proximal part of the nail) and the nail bed penetrate deep into the dermis and lie in the proximity of the distal interphalangeal joint. Further, the underlying dermis of the nail plate blends into the periosteum of the distal phalanx, thus firmly attaching the nail to the phalanx. (20X)

Eponychium

Epidermis

Nail groove

Nail plate

Nail bed

Nail root

FIGURE 12.22

LM of a sagittal section of a fetal finger at a higher magnification to show the relationship between the distal phalanx and the specialized appendage, the nail. In this micrograph the highly keratinized free edge of the nail, the eponychium, the epidermis overlying the nail, the nail fold, and skin beneath the free surface of the nail, the hyponychium, can be identified. (100X)

Cardiovascular System

The function of the **cardiovascular system** is to circulate the blood throughout the organ tissues of the body. Therefore, it provides for nutrition and **hormone** delivery, gas exchange, and the removal of waste from the tissues.

Anatomically, the cardiovascular system is organized on both a gross and a micromorphological level. At the gross level, the system consists of the **heart** and **vasculature**. At the microscopic level are tissues of several types that contribute to the system's histological appearance and organization.

The vasculature of the cardiovascular system is arranged in three fairly distinct layers or **tunics:** the outer layer or **tunica adventitia**, the middle layer or **tunica media**, and the inner layer or **tunica intima**. Each tunic has a preponderance of a particular type of tissue. The mass of the tissue within the tunic depends on whether the vasculature is **venous** or **arterial**.

Arterial Vasculature

Arteries are classified as **elastic arteries** (large), **muscular arteries** (medium), and **arterioles** (smallest arteries). Elastic arteries are found closer to the heart than either muscular arteries or arterioles. The anatomic progression, from proximal to distal, results in a morphological change from elastic arteries to arterioles.

Elastic arteries have three tunics: an **adventitia**, a **media**, and an **intima**. The tunica intima consists of two tissues: epithelium and connective. The intimal epithelium is a simple squamous epithelium called **endothelium**. The connective tissue component of the intima consists predominantly of elastic fibers, some forming incomplete laminae and others forming the internal **elastic lamina**. Along with the rest of the connective tissue matrix of elastic fibers, and **amorphous** substance, there is a very small amount of collagen fibers.

Certain types of cells, such as **fibroblasts**, are also found in conjunction with the connective tissue component of the intima. The tunica intima is arranged in layers, with the endothelium forming one layer and the subendothelium of connective tissue forming another layer. The internal elastic lamina forms a boundary of demarcation between the tunica intima and the tunica media.

The tunica media of elastic arteries is wider than the intima and is predominantly elastic tissue arranged concentrically in the form of **fenestrated laminae**. Cell types include fibroblasts, mesenchymal cells, and smooth muscle cells. Interspersed among the elastic laminae is a small amount of collagen fibers and amorphous ground substance. Like the internal elastic lamina of the tunica intima, the tunica media also has a well-organized array of fibroelastic fibers called the **external elastic** lamina. The lamina serves as a boundary between the media and the tunica adventitia.

The tunica adventitia of elastic arteries is thinner than the tunica media. The adventitia is composed of mixed irregular elastic and collagen tissues concentrically arranged superiorly to the external elastic lamina of the tunica media. Within the adventitia is the **vasa vasorum**, which consists of the blood vessels to the body wall of the artery. **Lymphatics** are also located within the adventitia.

Muscular arteries The muscular artery tunics vary somewhat in terms of thickness and composition. The tunica intima is thinner, and the tunica adventitia is thicker, with respect to its own media in muscular arteries. The media of muscular arteries consists of **concentrically** arranged **smooth muscle fibers**. Elastic fibers are found intercellularly within the tunica media. Muscular arteries also have both internal and external elastic laminae, which serve as boundaries between intima and media and between media and adventitia, respectively. The internal elastic media may be bifid or split in some arteries. The adventitia of muscular arteries also contains the **vasa vasorum** and **lymphatics**. Nerve tissue may also be seen traversing the adventitia to innervate the muscle.

Arterioles The tunica intima of arterioles consists of an **endothelium** and an internal elastic membrane. The tunica media is relatively thick, with little connective tissue and few smooth muscle cells. The subendothelial layer is lacking in arterioles. The internal elastic membrane vanishes as the arterioles become smaller. The external elastic lamina is undifferentiated, and the adventitia is thin and relatively small in relation to the other tunics.

Capillaries are simpler blood vessels. They are predominantly made of a **tunica intima** and a fine **tunica media**, but a nearly nonexistent **tunica adventitia**. The intima of capillaries consists of a one-cell **endothelium** and the underlying **basal lamina**. No true subendothelium is found, and capillaries lack an internal elastic lamina. Capillaries exhibit variability within the tunica intima depending on the location. The variability found within the endothelium layer is the basis of classification of capillary types.

Venous Vasculature

The **venous** vasculature follows the same morphological arrangement as do the arteries. The walls of veins are thinner as a result of reduction in the mass of medial components. Venous vasculature lacks a distinct **external elastic** lamina, and the internal elastic lamina is less obvious in most veins than in arteries. In general, there is more variability in the venous vasculature. The veins may be classified as **venules, small** and **medium** veins, and **large** veins. On the basis of the size and thickness of the body wall, veins are small, medium, or large. The small and medium veins are grouped together. The predominant tissue type varies among the venous vessels, but is similar to that in arterial vasculature.

Venules have a tunica intima that consists of **endothelium** and a **basal lamina**, but lack a true **lamina propria** or subendothelial region within the intima. The tunica media of small venules is essentially absent but may consist of **pericyte cells**. The pericyte cells are generally in contact with the basal lamina and the **endothelium**. As venules increase in size, the pericyte cells give way to an incomplete tunica media of scattered concentrically arranged **smooth muscle**. The tunica media becomes more pronounced as the vessels increase in size. The tunica **adventitia** is relatively thin and consists of collagen with a few interspersed **fibroblasts**.

Small and medium veins The tunica intima of small and medium veins consists of an endothelium with its basal lamina. In medium veins, subendothelial connective tissue forms the **lamina propria.** An ill-defined internal elastic lamina may be visible in the venous intima. The tunica media of small and medium veins is relatively thin, with less smooth muscle and more **collagen** than elastin. The tunica adventitia of these veins is composed of collagenous connective tissue with **elastic fibers, fibroblasts,** and **smooth** muscle cells. The adventitia is thicker in medium veins than the media and forms the bulk of the body wall. Like arteries, the **vasa vasorum** of veins are found in the adventitia.

Large veins There is more collagen relative to elastin in large veins, and the tunica intima consists of an **endothelium,** its basal lamina, and a lamina propria. The subendothelial layer is more prominent in the larger veins. The internal elastic lamina is more organized and defined within the intima of large veins. The tunica media is thin and consists of concentrically arranged **smooth muscle, collagen,** and **elastic fibers.** The tunica intima composes the mass of the vessel wall. It contains collagenous connective tissue loosely arranged and interspersed among bundles of **collagen fibers** and longitudinally arranged **fascicles** of smooth muscle. The adventitia of large veins contains the **vasa vasorum,** which may extend toward the intimal layer.

Heart

The **heart** is a four-chambered valvular structure that is similar in micromorphology to the **vasculature.** The body wall of the heart has three distinct layers: an **endocardium,** a **myocardium,** and an **epicardium.**

The **endocardium** of the heart is continuous with the intima of the vasculature. The endocardium is composed of **squamous** endothelium, its basal lamina, and a subendothelial layer of fine collagenous connective tissue. Below the subendothelium is the subendocardium, which consists of densely arranged connective tissue and smooth muscle fibers. Some **adipose** tissue may also be found deeper in the subendocardium. The specialized conducting tissue and subendocardial blood vessels are located in the subendocardium.

The **myocardium** is composed of striated cardiac muscle. The thickness of the myocardium varies with the location. The myocardium of the ventricles is many times thicker than that of the **atria.** The **ventricular** myocardium consists of three layers of interdigitating **muscle:** two outer **oblique layers,** and an inner **circular layer.** Throughout the myocardium are interspersed **elastic** fibers that are continuous with the fibers of the **endocardium** and the **epicardium.** Dense **collagenous** connective tissue is also found in the myocardium; this tissue comprises the **cardiac skeleton.**

The heart has a unique **conducting system** of specialized cardiac tissue. The system consists of the **SA node,** the **AV node,** and the **AV bundle.** The **SA node** is in the **epicardium** of the **right atrium** and is composed of a group of specialized **cardiac cells,** intercellularly arranged **elastic** connective tissue, and innervating **nervous** tissue. The **AV node** is a group of specialized cells located within the **subendocardial** layer at the junction of the **right atrium** and the **right ventricle.** The AV node projects fibers called the **AV bundle** into the **interventricular septum.** The AV node also receives innervation through fibers from the SA node.

The **epicardium** is composed of a **mesothelium** and an underlying **subepicardium** of connective tissue. The mesothelium is **squamous** to **cuboidal simple epithelium.** The subepicardium is composed of fine **loose areolar** connective tissue and **elastic fibers.** The subepicardial region contains the **coronary arteries,** the **veins,** and the **lymphatics.**

The heart has four **chambers** with **valves,** which delineate the boundaries between the chambers and designate outflow tracts. There are four valves associated with the heart: the **tricuspid,** the **bicuspid** or **mitral,** the **pulmonary,** and the **aortic semilunar valves.**

Cardiovascular System—Heart

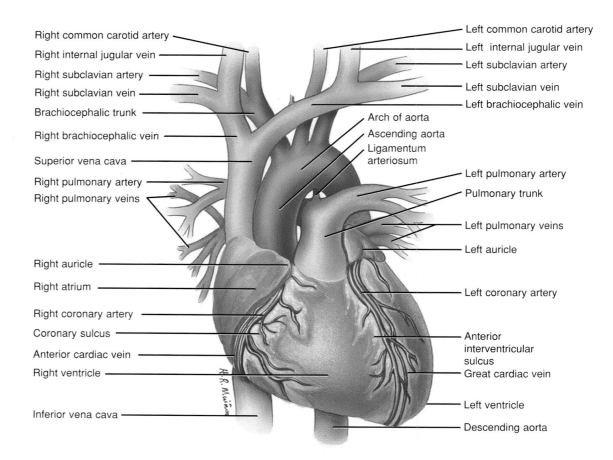

Right common carotid artery

Right internal jugular vein

Right subclavian artery

Right subclavian vein

Brachiocephalic trunk

Right brachiocephalic vein

Superior vena cava

Right pulmonary artery

Right pulmonary veins

Right auricle

Right atrium

Right coronary artery

Coronary sulcus

Anterior cardiac vein

Right ventricle

Inferior vena cava

Left common carotid artery

Left internal jugular vein

Left subclavian artery

Left subclavian vein

Left brachiocephalic vein

Arch of aorta

Ascending aorta

Ligamentum
arteriosum

Left pulmonary artery

Pulmonary trunk

Left pulmonary veins

Left auricle

Left coronary artery

Anterior
interventricular
sulcus

Great cardiac vein

Left ventricle

Descending aorta

FIGURE 13.1

Diagram of an anterior view of the heart and supporting blood vessels
associated with the cardiovascular system. Also shown are coronary
blood vessels and their distribution in the cardiac tissue.

FIGURE 13.2
Light micrograph (LM) of a cross section through a fetal heart with developing embryonic myocardium. The micrograph shows the myocardium and the epicardium with a blood vessel. The nuclei of the cardiac muscle are prominent. Some elastic fibers are visible. The epicardium has simple squamous cells with an underlying subepicardium of loose areolar and adipose connective tissue. (100X)

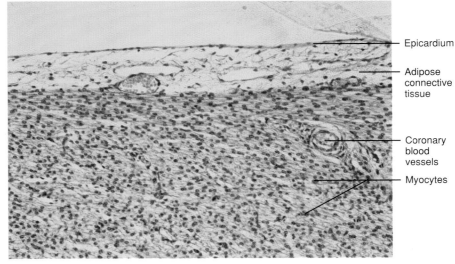

Epicardium

Adipose connective tissue

Coronary blood vessels

Myocytes

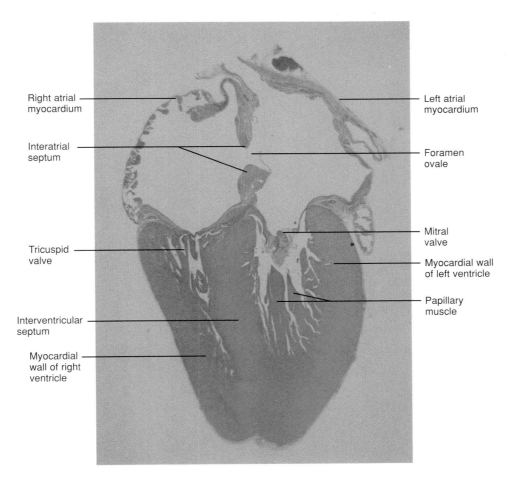

Right atrial myocardium

Interatrial septum

Tricuspid valve

Interventricular septum

Myocardial wall of right ventricle

Left atrial myocardium

Foramen ovale

Mitral valve

Myocardial wall of left ventricle

Papillary muscle

FIGURE 13.3
LM of a longitudinal section through a fetal heart, showing the four chambers. Papillary muscle within the ventricle can be identified. The interventricular septum separates the left and right ventricles. The interatrial septum with a **foramen ovale** can be seen in the atrial area. Atrial body walls are much thinner than ventricular walls. (1X)

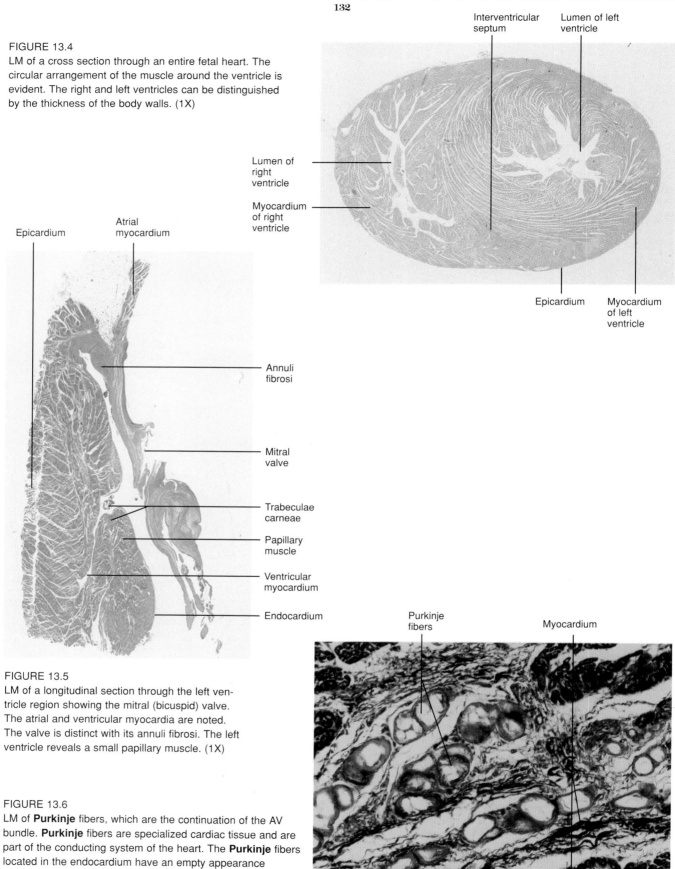

FIGURE 13.4
LM of a cross section through an entire fetal heart. The circular arrangement of the muscle around the ventricle is evident. The right and left ventricles can be distinguished by the thickness of the body walls. (1X)

Interventricular septum

Lumen of left ventricle

Lumen of right ventricle

Myocardium of right ventricle

Epicardium

Myocardium of left ventricle

Epicardium

Atrial myocardium

Annuli fibrosi

Mitral valve

Trabeculae carneae

Papillary muscle

Ventricular myocardium

Endocardium

FIGURE 13.5
LM of a longitudinal section through the left ventricle region showing the mitral (bicuspid) valve. The atrial and ventricular myocardia are noted. The valve is distinct with its annuli fibrosi. The left ventricle reveals a small papillary muscle. (1X)

FIGURE 13.6
LM of **Purkinje** fibers, which are the continuation of the AV bundle. **Purkinje** fibers are specialized cardiac tissue and are part of the conducting system of the heart. The **Purkinje** fibers located in the endocardium have an empty appearance resulting from a lack of contractile elements within the cytoplasm. (200X)

Purkinje fibers

Myocardium

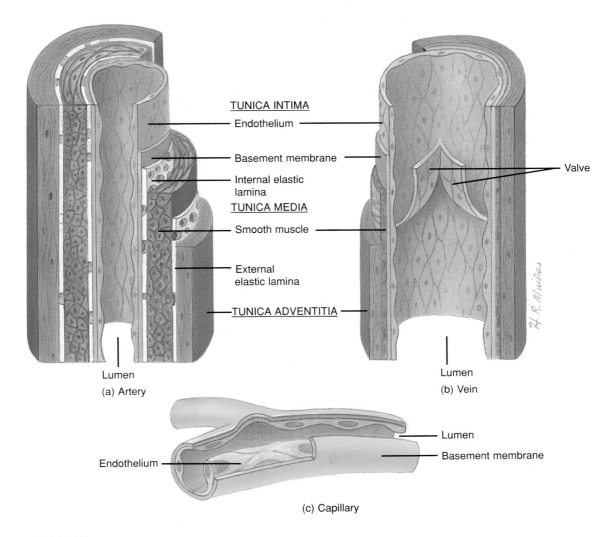

TUNICA INTIMA
Endothelium
Basement membrane
Internal elastic lamina
Valve
TUNICA MEDIA
Smooth muscle
External elastic lamina
TUNICA ADVENTITIA
Lumen
(a) Artery
Lumen
(b) Vein

Lumen
Basement membrane
Endothelium
(c) Capillary

FIGURE 13.7
Diagrams of (a) an artery, (b) a vein, and (c) a capillary, illustrating the morphological differences in their body walls.

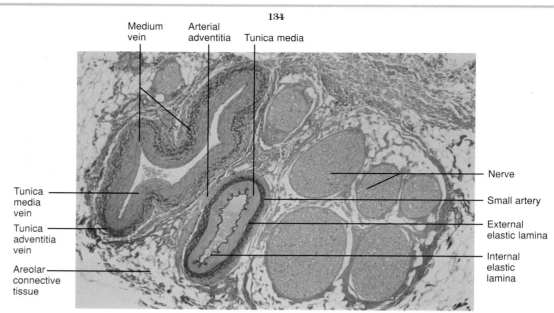

Medium vein   Arterial adventitia   Tunica media

Tunica media vein

Tunica adventitia vein

Areolar connective tissue

Nerve

Small artery

External elastic lamina

Internal elastic lamina

FIGURE 13.8
LM of a cross section through a neurovascular bundle. The micrograph reveals a medium vein, a small artery, and several small nerves and lymphatics. The surrounding areolar connective tissue is also present. The artery shows an internal elastic lamina, a prominent tunica media, a well-defined external elastic lamina, and a fairly distinct adventitia. Blood can be seen within the lumina. The medium vein, larger than the artery, lacks a distinct internal elastic lamina. The media in the body wall has an external elastic lamina that is less organized than the arterial counterpart. The adventitia of the vein is prominent and extends into the surrounding connective tissue. (100X)

FIGURE 13.9
LM of a transverse section through the body wall of the aorta, showing a thick tunica intima and a gradual change to tunica media. The tunica media has multiple laminae of elastic fiber. Internal and external elastic laminae are not prominent. Spaces between the elastic laminae contain some smooth muscle cells and collagen fibers. The vasa vasorum is located within the adventitia. (100X)

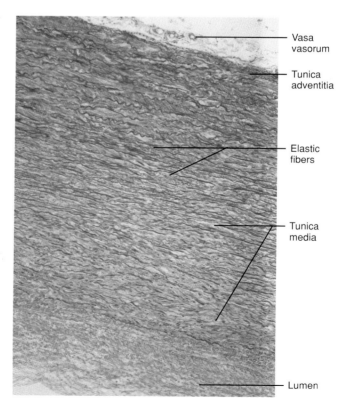

Vasa vasorum

Tunica adventitia

Elastic fibers

Tunica media

Lumen

FIGURE 13.10
LM of a transverse section through the tunica media of the aorta, showing prominent elastic fibers with interspersed smooth muscle fiber, fibroblasts, and amorphous substance. (400X)

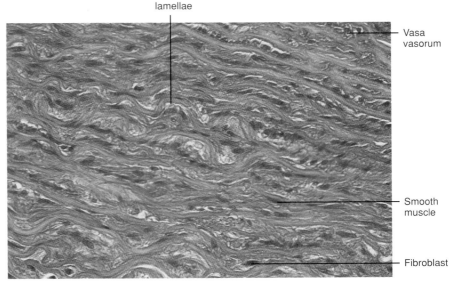

Elastic lamellae

Vasa vasorum

Smooth muscle

Fibroblast

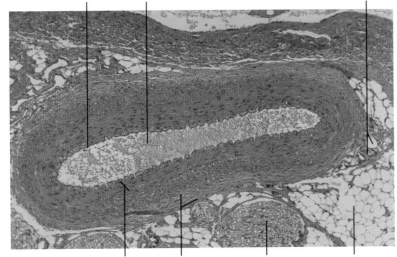

Internal elastic lamina

Lumen of muscular artery

Lymph vessel and vasa vasorum

Tunica media

Tunica adventitia

Nerve

Adipose tissue

FIGURE 13.11
LM of a cross section through a muscular artery. The internal elastic lamina is distinct and forms a boundary between the thin tunica intima and the thick, smooth muscle tunica media. The external elastic lamina is present but less distinctly demarcated than the internal laminae. The adventitia is prominent but contributes less to the vascular wall than does the media. The adventitia contains vasa vasorum and lymphatics, as shown to the right of the vessel in the micrograph. Adipose tissue surrounds the adventitia. Nerve tissue is also located below the vessel. (100X)

FIGURE 13.12
LM of a cross section of a muscular artery. The lumen contains numerous red blood cells (RBC). The intima is demarcated by the nuclei of the endothelium and the internal elastic membrane. The media is prominent in the muscular artery, and smooth muscle nuclei are visible throughout the tunica media. The external elastic laminae can be identified by wavy fibers of elastin. A prominent adventitia covers the blood vessel. **Verhoeff's** stain. (200X)

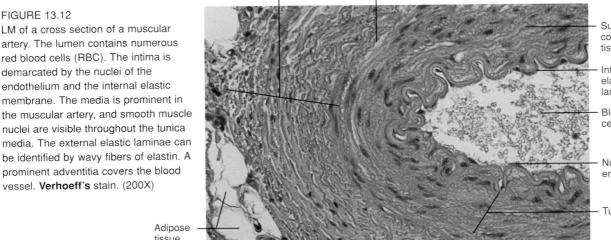

Tunica adventitia

External elastic lamina

Subendothelial connective tissue

Internal elastic lamina

Blood cells

Nuclei of endothelium

Tunica media

Adipose tissue

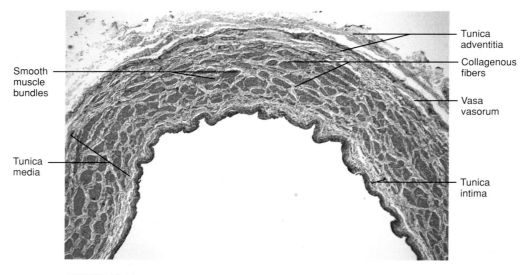

Smooth
muscle
bundles

Tunica
media

Tunica
adventitia

Collagenous
fibers

Vasa
vasorum

Tunica
intima

FIGURE 13.13

LM of a cross section of a large vein. The adventitia and tunica media are the most prominent structures, with collagenous and elastic fibers and smooth muscle bundles. The tunica adventitia has longitudinal bundles of smooth muscle interspersed among the collagen fibers. The tunica media is prominent, and the tunica intima can be identified by the elastic lamina. The subendothelial layer, the lamina propria (not shown), is thicker than other veins. (40X)

FIGURE 13.14

LM of a cross section of a large vein, showing prominent tunica adventitia and tunica media with elastic fibers, longitudinal bundles of smooth muscles, and large concentrations of collagen fibers. The media is shown in the right lower corner along with the intimal layer and lumen. (200X)

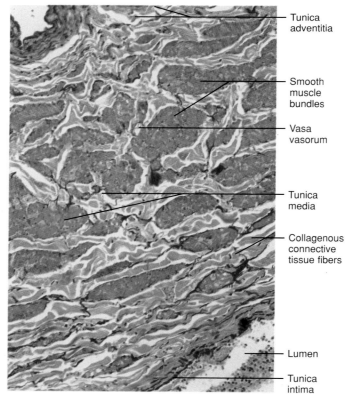

Tunica
adventitia

Smooth
muscle
bundles

Vasa
vasorum

Tunica
media

Collagenous
connective
tissue fibers

Lumen

Tunica
intima

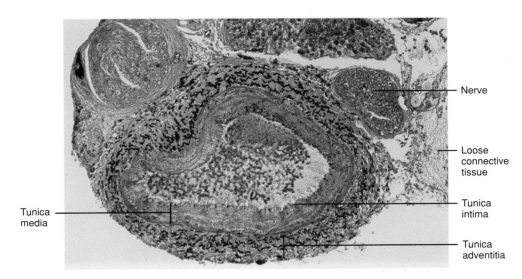

FIGURE 13.15
LM of a cross section of a medium vein. The tunica intima is thin, and is lined by a thick internal elastic lamina. The media of veins have more collagen fibers and less elastic fibers than medium muscular arteries. The media is thinner than that of medium arteries. The adventitia is prominent and is thicker than the tunica media. It contains vasa vasorum, which are generally more prominent in large veins than in arteries. **Verhoeff's** stain. (200X)

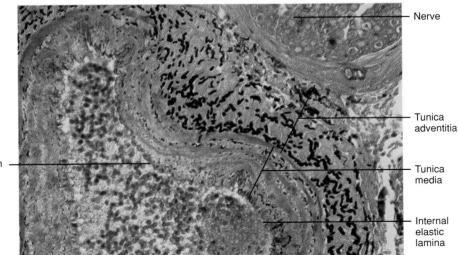

FIGURE 13.16
LM of a cross section of a medium vein. As shown in this micrograph, the intima is thin, the internal elastic lamina is prominent, and the endothelium squamous cell nuclei are distinct. The tunica media lies above the internal elastic lamina, and reveals both elastic fibers and dense collagen fibers. Smooth muscle cells are located between the fibers, with some layers of collagen fibers concentrically arranged. The adventitia of veins is much larger than that of arteries. **Verhoeff's** stain. (400X)

FIGURE 13.17

LM of a cross section of a medium vein. Prominent in this micrograph is the adventitia of connective tissue. The external elastic lamina blends with the adventitia. The media is composed of more collagen fibers than elastic fibers. The medial muscle fibers are less dense than their muscular artery counterparts. (200X)

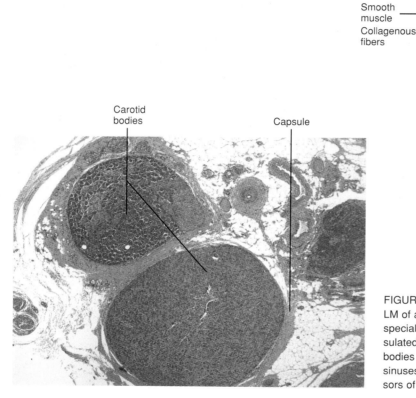

Carotid bodies — Capsule

FIGURE 13.18

LM of a section through carotid bodies. Carotid bodies are specialized masses of epithelial (glomus) cells that are encapsulated. They are innervated with nerve endings. The carotid bodies are located in the walls of the aorta and carotid sinuses, and act as chemoreceptors, pH sensors, and sensors of blood pressure changes in the arterial system. (40X)

Elastic fibers

Tunica adventitia

External elastic lamina

Smooth muscle

Collagenous fibers

Tunica intima — Endothelium

FIGURE 13.19

LM of a cross section through a glomus of thick skin. Glomuses are arteriovenous anastomoses common in the dermis of thick skin. The smooth muscle cells of the arterioles have hypertrophied, and appear similar to epithelioid cells. Capillaries and nerves may also be present in a glomus. A dense connective tissue sheath in the form of a capsule surrounds the glomus. (400X)

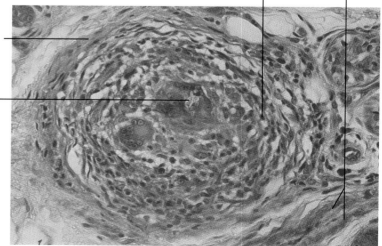

Glomus: media of epithelioid muscle cells

Connective tissue of dermis

Collagenous sheath around the glomus

Lumen of a blood vessel

The Lymphatic System

CHAPTER

14

The **lymphatic system** functions as part of both the **immune** system and the **circulatory** system. As part of the immune system, it functions in the proliferation, development, and storage of immune cells and as a site for immune reactions to occur. As part of the circulatory system, it functions in the removal of excess interstitial fluid and contributes to fluid **homeostasis.**

The lymphatic system is organized on a gross anatomical level and on a microanatomical level. It consists of several cellular components organized to form solid, vascular, or mixed organs. On the gross level, the anatomical arrangement of the lymphatic system is diffused and consists of **lymph nodules, lymph nodes, tonsils, thymus, spleen, lymphatic vessels,** and **Peyer's patches.**

The **lymph nodules** are well-defined aggregates of lymphoid tissue within the mucosae of certain organs. Nodules are found in the **lamina propria** of the **gastrointestinal tract,** the **tongue,** the **tonsils,** the **lymph nodes,** and the tissue of the **respiratory** system. Lymph nodules are not encapsulated.

The **lymph nodes** are more complex than the nodules. A lymph node may have several lymph nodules in its **stroma** of connective tissue. Surrounding

the node is a thin **connective tissue capsule.** The capsule projects **trabeculae** from its inner surface into the underlying lymphoid tissue, forming boundaries for individual lymph nodules. The capsular trabeculae serve as scaffolding for interwoven **reticular fibers.** In this way the trabeculae and reticular fibers contribute to the **stroma** of the lymph node. The cortical stroma of the lymph node is infiltrated with aggregates of **lymphocytes.** The lymphocyte aggregates and the density of the **reticulate fibers** contribute to the separate **medullary** and **cortical** regions within the lymph node. Lymphocytes are confined to the cortex, and reticulate fibers are concentrated in the medullary region.

The **lymph node cortex** is the region beneath the capsule and is divided into an **outer cortex** and an inner **paracortex.** The outer cortex is composed of **lymphoid nodules.** Within the lymphoid nodules are **germinal centers** composed of proliferating **B-lymphocytes.** The paracortex consists of B- and T-lymphocytes.

The **medulla** of the **lymph node** lies below the cortex, where the lymphoid tissue is less densely arranged. The cells of the medulla are organized to form **medullary cords.** The cords are composed of lymphocytes, plasma cells, and macrophages. Also present in the medulla are reticular fibers that interconnect with the trabeculae and the capsule. Lymph nodes also possess sinuses, which are composed of attenuated endothelium. The sinuses are continuous with the lymphatic vessels and are anatomically named by their location.

Three types of tonsils can be identified. Based on their anatomic location, tonsils are lingual (at the base of the tongue), palatine (two, located between the glossopalatine and pharyngopalatine arches), and a single pharyngeal tonsil (located along the midline of the posterior wall of the nasopharynx). Tonsils in general are lymphoid tissue with a surrounding connective tissue capsule. They are covered with a well-defined epithelium. The location of the tonsil dictates the epithelial type. The lingual and the palatine tonsils are covered by stratified squamous epithelium, and the pharyngeal tonsil is covered with pseudostratified ciliated columnar epithelium. Like other lymphoid organs, tonsils have lymphoid nodules with active germinal centers.

The thymus is an encapsulated organ consisting of two lobes. The lobes are divided into lobules by connective tissue septa. Similar to lymph nodes, the thymic lobules have both medullary and cortical regions. The cortex of the thymus is relatively denser than the medulla. The cortical cells consist of small, medium, and large lymphocytes. The medulla of the thymus is less densely packed with lymphocytes. Unique to the thymic medulla are structures known as Hassall's corpuscles. Hassall's corpuscles are reticular epithelial cells wrapped around themselves to form laminated structures. The function of these cells is unknown.

The spleen, like the lymph nodes, tonsils, and thymus, is an encapsulated organ. The capsule is composed of dense connective tissue covered by squamous mesothelium. Also found in the capsule are elastin and smooth muscle. The capsule extends deep into the splenic substance to form trabeculae. Reticular fibers are interwoven between the trabeculae and concentrations of cells. Collectively, the trabeculae and the cells form the stroma of the spleen. The stroma functions as a skeleton and a support network for the deeper structures within the spleen. These deeper structures are arteries, veins, and the cellular parenchyma.

The splenic parenchyma consists of red and white pulp. The red pulp is associated with the venous components of the spleen. It consists of a network of venous sinuses and an array of reticuloendothelial cells and their components arranged to form Billroth's or splenic cords. The white pulp is associated with the arterial system within the spleen. It consists of lymphoid tissue adhering to the adventitial layers of the arteries. The lymphoid tissue is arranged into sheaths and gradually merges into lymphoid splenic nodules.

The lymphatic vessels consist of blind lymphatic capillaries, collecting vessels, and lymphatic trunks. The morphological structures are similar to the venous vasculature and blood capillaries. The lymph capillaries are composed of an endothelium and its basal lamina with no true media or adventitia. The endothelium is covered with an external layer of connective tissue. The lymphatic collecting vessels are analogous to veins, and have three tunics. The tunica intima consists of an endothelium, its basal lamina, and an underlying lamina propria. The tunica media consists of smooth muscle with interspersed elastic fibers. The tunica adventitia is the largest layer and consists of collagen, elastin, and smooth muscle. The lymphatic vessels also possess unidirectional valves and a vasa vasorum, similar to their vascular counterparts, the veins. The lymphatic trunks are the largest of the lymphatic vessels and are comparable to large veins. The tunica intima consists of an endothelium, a basal lamina, and a lamina propria of fibroelastic tissue. The tunica media is predominantly smooth muscle with some collagen and elastic fibers. The tunica media contributes to the thickness of the vessel. The tunica adventitia of the lymphatic trunk is predominantly collagenous connective tissue with some elastic fibers and smooth muscle interspersed within the layer. The adventitia is ill-defined and blends with the surrounding connective tissue.

The Peyer's patches are aggregates of lymphatic nodules found in the appendix and the lower half of the small intestine. The patches consist of many secondary nodules forming large oval structures that lie below the epithelium. Peyer's patches are more prevalent in children, but decrease in number in adults.

Lymph Node

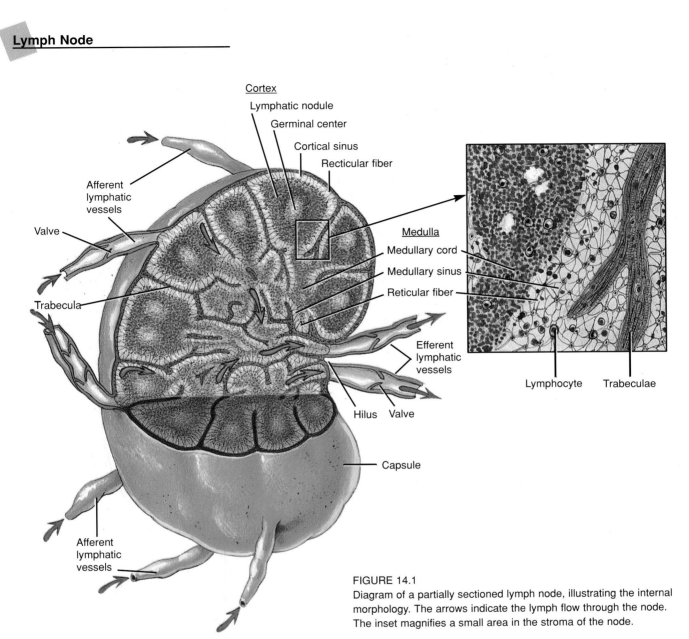

Cortex
Lymphatic nodule
Germinal center
Cortical sinus
Recticular fiber

Afferent
lymphatic
vessels

Valve

Trabecula

Medulla
Medullary cord
Medullary sinus
Reticular fiber

Efferent
lymphatic
vessels

Hilus   Valve

Capsule

Afferent
lymphatic
vessels

Lymphocyte   Trabeculae

FIGURE 14.1
Diagram of a partially sectioned lymph node, illustrating the internal
morphology. The arrows indicate the lymph flow through the node.
The inset magnifies a small area in the stroma of the node.

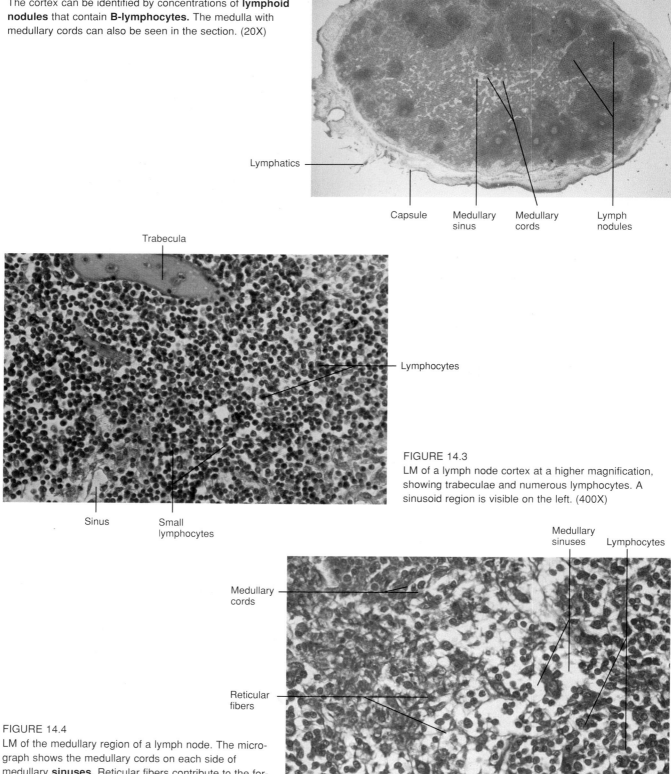

FIGURE 14.2
Light micrograph (LM) of a lymph node, showing the connective tissue capsule with subcapsular and medullary sinuses. The cortex can be identified by concentrations of **lymphoid nodules** that contain **B-lymphocytes.** The medulla with medullary cords can also be seen in the section. (20X)

Subcapsular sinus

Lymphatics

Capsule

Medullary sinus

Medullary cords

Lymph nodules

Trabecula

Lymphocytes

FIGURE 14.3
LM of a lymph node cortex at a higher magnification, showing trabeculae and numerous lymphocytes. A sinusoid region is visible on the left. (400X)

Sinus

Small lymphocytes

Medullary sinuses

Lymphocytes

Medullary cords

Reticular fibers

FIGURE 14.4
LM of the medullary region of a lymph node. The micrograph shows the medullary cords on each side of medullary **sinuses.** Reticular fibers contribute to the formation of the stroma. Numerous lymphocytes can also be seen in the stroma. (400X)

FIGURE 14.5
LM of a lymph node at a higher magnification. The micrograph shows small,
medium, and large lymphocytes, and
macrophages forming the majority of the
parenchyma. (1000X)

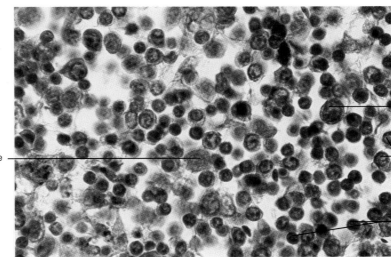

Large
lymphocyte

Medium
lymphocytes

Small
lymphocytes

Macrophage

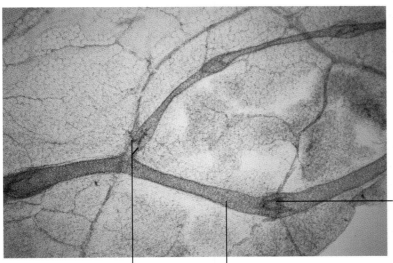

Valve

Valve
leaflets

Lymphatic
vessel

FIGURE 14.6
LM of a lymphatic vessel and blood capillaries. The
micrograph shows valves in the lymph vessel and their
leaflets. The lumen of the vessel is expanded in diameter at the locations of the valves. (40X)

Lymph
vessels

Endothelium

Valve

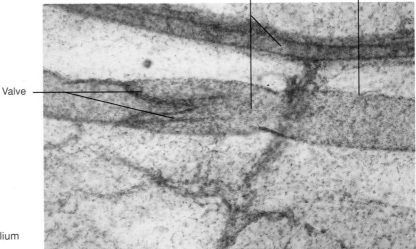

FIGURE 14.7
LM of a lymphatic vessel at a higher magnification,
revealing the leaflets of the lymphatic valve. Endothelium
lines the lymph vessel. (100X)

FIGURE 14.8
LM of a palatine tonsil. The micrograph
reveals several lymphoid nodules with
numerous germinal centers. The nodules
are covered with stratified squamous epithelium, invaginating at places into tonsillar
crypts. (100X)

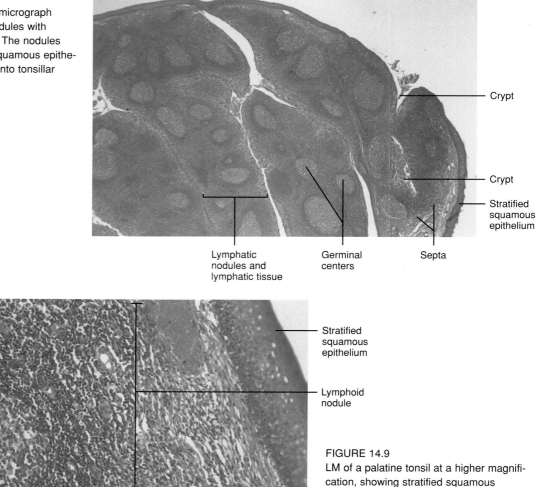

Crypt

Crypt

Stratified
squamous
epithelium

Lymphatic
nodules and
lymphatic tissue

Germinal
centers

Septa

Stratified
squamous
epithelium

Lymphoid
nodule

Germinal
center

Vessels

FIGURE 14.9
LM of a palatine tonsil at a higher magnification, showing stratified squamous
epithelium covering a lymphoid nodule. A
germinal center can be seen within the
nodule. Vessels are apparent in the connective tissue. (200X)

FIGURE 14.10
LM of a lingual tonsil. Stratified squamous epithelium
covers the lymphoid tissue. Germinal centers of lymphoid nodules are prominent below the epithelium.
Deep in the lymphoid tissue is the stroma of connective
tissue, glossal muscle, some glandular tissue, and adipose tissue. (20X)

Mucous
gland

Adipose
tissue

Lymphatic tissue
with lymphoid
nodules

Germinal
center

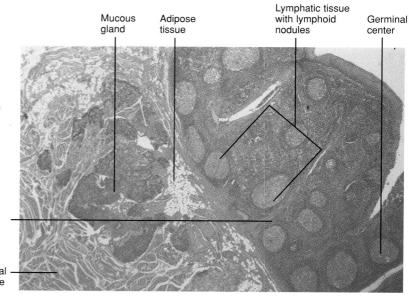

Septa

Glossal
muscle

FIGURE 14.11
LM of a lingual tonsil at a higher magnification, revealing lymphoid nodules with germinal centers, septa of connective tissue, adipose tissue, and numerous glands deep in the stroma of the tonsil. (40X)

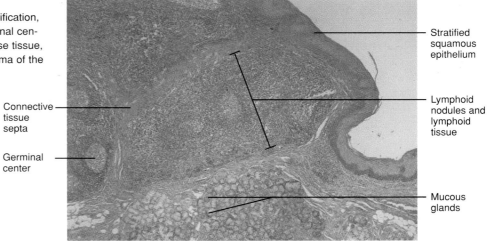

Stratified squamous epithelium

Lymphoid nodules and lymphoid tissue

Connective tissue septa

Germinal center

Mucous glands

Tonsillar crypt

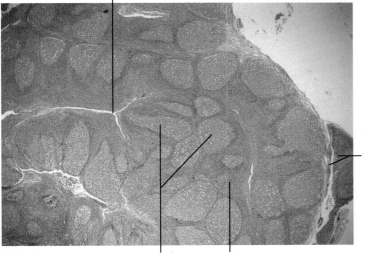

FIGURE 14.12
LM of a pharyngeal tonsil, showing numerous lymphoid nodules and tonsillar crypts. The pharyngeal tonsil is the smallest of the three types of tonsils and is lined by pseudostratified ciliated columnar epithelium. (200X)

Epithelium

Lymphoid nodules

Diffuse lymphoid tissue

Goblet cells   Cilia

Ciliated pseudostratified columnar epithelium

Vessel

Lymphoid tissue

FIGURE 14.13
LM of a pharyngeal tonsil at a higher magnification. The tonsil is lined by pseudostratified ciliated columnar epithelium with interspersed goblet cells. Loose areolar connective tissue with diffused lymphocytes can be seen below the epithelium. A lymphatic vessel is also visible. (400X)

Epithelium

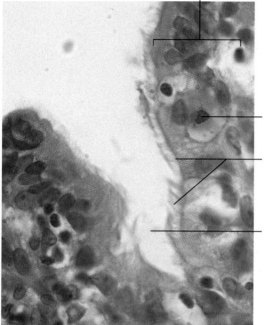

Nucleus

Cilia

Crypt

FIGURE 14.14
LM of a pharyngeal tonsil, showing a predominance of pseudostratified ciliated columnar epithelium and some lymphocytes below the epithelium. (1000X)

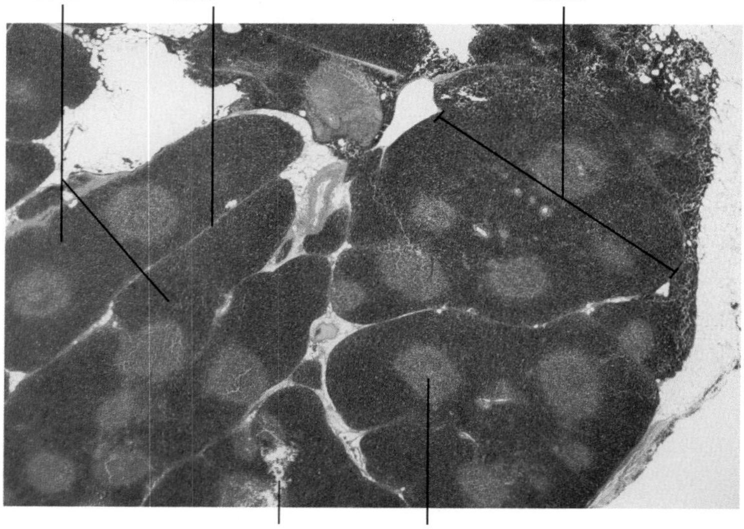

Cortex — Septa with connective tissue — Lobule

Trabeculae — Medulla

FIGURE 14.15
LM of thymus tissue in cross section. A collagenous capsule surrounds the thymus. The thymic lobules are separated by a thin, connective-tissue-filled septa. The cortex of the lobule stains dark blue and consists of small, medium, and large lymphocytes. The medulla is stained lighter and is more centrally located. (20X)

Lymphocytes — Hassall's corpuscle

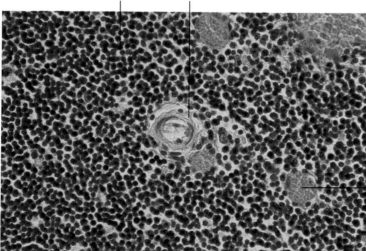

Red blood cells

FIGURE 14.16
LM of a thymic lobule in cross section. The section shows the medulla of the lobule. The medulla has a cellular density less than that of the cortex. There is a predominance of small lymphocytes. Other cells, such as macrophages, plasma cells, and mast cells, are also found in the medulla. A **Hassall's** corpuscle is visible in the section. (400X)

FIGURE 14.17
LM of a thymus, showing numerous medullary lymphocytes and **Hassall's** corpuscles. The function of **Hassall's** corpuscles in the thymus is uncertain. (400X)

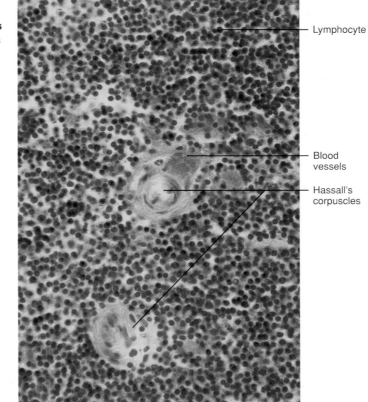

Lymphocyte

Blood vessels

Hassall's corpuscles

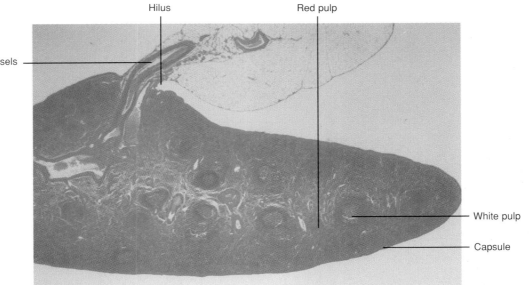

Hilus

Red pulp

Vessels

White pulp

Capsule

FIGURE 14.18
LM of the spleen at low magnification. The spleen is surrounded by a thin fibrous capsule. The fibrous capsule blends with the internal connective tissue trabeculae that surround the large blood vessels. The fibers traverse the entire spleen, forming a connective tissue network. Splenic vessels are penetrating the capsule. Both red and white pulp can be identified in the section. (20X)

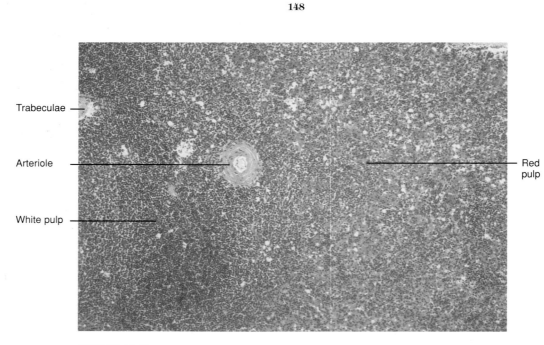

Trabeculae

Arteriole

White pulp

Red pulp

FIGURE 14.19
LM of the spleen at a higher magnification, displaying areas of red and white pulp. Also visible are a central arteriole, sinusoids, and trabeculae in cross section. (100X)

Small lymphocytes

Splenic cords with lymphocytes

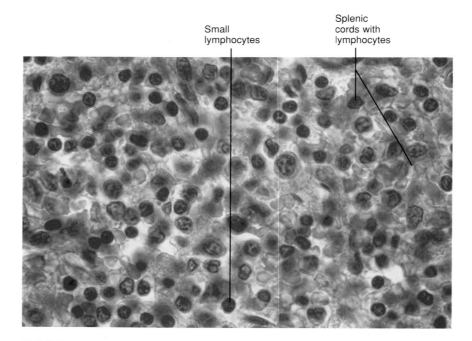

FIGURE 14.20
LM of the spleen at a still higher magnification. The sinuses exhibit a variety of cells, including lymphocytes, erythrocytes, smooth muscle cells, macrophage cells, and endothelial cells, which line the sinusoids of the spleen. (1000X)

Digestive System

The large and fairly extensive **digestive system** begins with the **oral cavity,** extends through the **thoracic, abdominal,** and **pelvic cavities,** and terminates with the **anus.** The system has two basic components: the **digestive system proper** or **alimentary tract** and its associated **glands** and **organs.**

The **digestive system proper** is organized into **tunics,** and its gross morphology is tubular in nature. It consists of several histological components with tissue variance within the tunics.

Overall, the organs of the digestive system are the **oral cavity** (including the **salivary glands, teeth,** and **tongue), oral pharynx, esophagus, stomach, duodenum, jejunum, ileum, colon, rectum, anus, pancreas, liver, gallbladder,** and the **mucosal glands** of the **alimentary tract.**

The alimentary tract begins with the oral cavity, which consists of the tongue, teeth, salivary glands, lips, and cheeks. The **oral cavity** begins with the lips, which are lined by **stratified squamous epithelium** with an underlying connective tissue. The **vermilion border** that separates each lip into the distinct "red lip" is nonkeratinized stratified squamous epithelium and lacks the typical skin appendages. The lamina propria of the red portion of the lip forms numerous **papillae.** The papillae are vascular with a prominent capillary network. The inner lining of the **lips** is also stratified squamous epithelium. Skeletal muscle **fascicles** insert into the connective tissue bundles of the lip. The submucosal region is below the lamina propria of the lips. This region has numerous **mucoserous** salivary glands.

The **gingiva** (lower) and buccal mucosa are composed of a nonkeratinized stratified squamous epithelium with an underlying lamina propria. The **buccal mucosa** has a deeper submucosa that contains numerous mucoserous glands, capillaries, and lymphatics. The gingival mucosa, unlike the buccal mucosa, lacks a submucosa; the lamina propria is attached to the alveolar periosteum.

Tongue The tongue is a relatively complex organ consisting of muscle fascicles arranged in three planes perpendicular to each other. The dorsal surface of the tongue is covered with a mucosa of keratinized stratified squamous epithelium. Below the epithelium of the mucosa lies the lamina propria, but the tongue, like the gingiva, lacks a submucosa. The mucosal epithelium is connected to the underlying muscle through intermeshing of the connective tissue of the lamina propria. The ventral surface of the tongue is continuous with the floor of the mouth through its **mucous membrane.** The mucous membrane of the ventral tongue is composed of a nonkeratinized stratified squamous epithelium with an underlying lamina propria. Below the mucosa is the submucosa.

The dorsal surface of the tongue is covered with several types of epithelial projections or **papillae: fungiform, filiform, foliate (rare),** and **circumvallate** papillae. **Taste buds** are specialized **chemoreceptors** that are located primarily in the circumvallate and foliate papillae. They are barrel-shaped structures composed of fusiform or curved **sustentacular** or **neuroepithelial** cells.

Glands   Several types of **salivary glands** are associated with the oral cavity. These **tuboalveolar compound glands** are considered serous, mucous, or mixed and are classified as minor or major. The minor salivary glands are intrinsic to the oral mucosa and are found in the submucosal region of the tissue. The major salivary glands are the **parotid, submandibular (submaxillary),** and **sublingual glands.** Major salivary glands lie outside the oral cavity but are connected by glandular ducts.

The **parotid** gland is the largest of the major salivary glands. It is a relatively large gland and has a fibrous connective tissue capsule with septa that divide the gland into lobes and lobules. The capsular septa, with other connective tissue elements, give rise to the stroma that supports the parenchyma and duct system of the gland. The parotid parenchyma consists of serous compound tuboalveolar glands. The cells are pyramidal in shape with a basophilic cytoplasm. Between these pyramidal epithelial cells are **myoepithelial** cells.

The parotid gland has a fairly extensive duct system beginning with the **intercalated duct.** This duct is composed of low cuboidal or squamous epithelium interspersed with myoepithelial cells. The intercalated duct gives rise to the **striated duct,** which consists of columnar cells with evaginations of the cell membrane, producing a striated appearance. Collectively, the intercalated and striated ducts form the **intralobular duct** system. The

striated duct is linked to **interlobular ducts.** These ducts are lined with columnar epithelium that becomes pseudostratified columnar epithelium and then striated columnar epithelium. The ducts then merge to form **Stensen's duct,** which is lined with stratified squamous epithelium.

The **submandibular gland (submaxillary gland)** is a **compound tuboalveolar gland** that also is enclosed in a connective tissue capsule. The capsular connective tissue extends into the parenchyma of the gland and divides the gland into lobes and lobules. The glandular parenchyma is comprised of mixed **serous** and **mucous** cells with a predominance of serous cell types. Its secretory units also contain myoepithelial cells. The submandibular gland also has a fairly extensive duct system including intercalated and striated ducts as well as a major excretory duct, **Wharton's duct.**

The **sublingual glands** are a composite of mixed **tuboalveolar glands.** The sublingual gland capsules are not as prominent, and the duct system is not as extensive, as in the parotid and submandibular glands. The parenchymal cells are predominantly serous, but myoepithelial cells are also present as part of the glandular tissue unit.

Teeth   The **teeth** contain both hard and soft tissue. The hard tissues of the teeth, which are arranged in layers, include the **cementum,** the **dentin,** and the **enamel.** The soft tissues are the **pulp,** the **periodontal membrane,** and the surrounding **gingiva.**

The dentin is a hard tissue and constitutes the bulk of the tooth. It is composed of **hydroxyapatite crystals** arranged on collagen fibers. The dentin appears striated with tubular structures coursing through it. These tubular structures are housings for the cytoplasmic projections of the **odontoblasts.** These cytoplasmic projections are called **Tome's fibers.** The dentin is covered at the free oral surface by enamel. Enamel is compressed into prisms and arranged perpendicularly to the dentin in spirals. The **cementum** lies between the dentin and the inner surface of the gingiva. It is an organic matrix similar to bone that is synthesized by **cementocytes.** The two types of cementum can be identified as cellular and acellular. **Sharpey's fibers,** which are also important in stabilizing teeth, are collagenous fibers that extend from the teeth and attach to the **alveolar bone** of the jaw.

In the center of the tooth is a chamber called the **pulp chamber.** The pulp is essentially collagen connective tissue with a gelatinous matrix. Cell types within the pulp include **fibroblasts, mesenchymal cells, macrophages,** and **lymphocytes.** Odontoblasts are found at the junction of the pulp cavity and the dentin. Blood vessels, nerves, and lymphatics are also present in the pulp.

The **periodontal ligament** consists of thick bundles of collagenous fiber that run between the cementum of the root and the alveolar bone of the jaw. Continuous with the periodontal ligament is the gingiva. Although not a part of the teeth, it contributes to the support of the teeth.

There are 20 **deciduous** teeth (baby or milk teeth) during childhood. The deciduous teeth are replaced by a permanent set of teeth beginning at age six. Around age 18, the third molar erupts, completing the set of 32 **permanent** teeth.

Gingiva The gingiva or gum is lined with keratinized stratified squamous epithelium. The gingiva surrounds each tooth and attaches to the crown of the tooth on one end and to the **periosteum** of the alveolar bone on the other end. The root of the tooth is thereby stabilized in the bony sulcus.

Esophagus The esophagus is a fairly straight, muscle-walled tubular structure. The mucosal lining is of nonkeratinizing stratified squamous epithelium. Also present in the body wall is the submucosa with esophageal gland, a muscularis with circularly and longitudinally arranged smooth muscle fibers, and an outer **adventitia.**

The **bolus** (mixture of food and saliva) passes from the oral cavity into a large opening, the **pharynx.** The bolus is channeled into the esophagus by the contraction of the pharynx muscle. The bolus is then systematically forced downward from the esophagus into the stomach by the **peristaltic** action of the esophagus.

Stomach The body wall of the stomach is divided into four layers: the **serosa** (the outermost), **muscularis, submucosa,** and **mucosa.** The mucosa is lined with columnar cells. As the bolus enters the stomach through the **cardiac sphincter,** it is quickly reacted upon by digestive enzymes and **hydrochloric acid** (HCl) that is secreted by the gastric glands located in the gastric pits. The **gastric glands** are composed of four types of cells: (1) **neck cells,** which secrete mucus that forms a protective layer on the gastric mucosa and the gastric pit; (2) **chief cells,** or **peptic cells,** which secrete the proteolytic enzyme **pepsin;** (3) **parietal cells,** or oxyntic cells, which secrete hydrochloric acid and an intrinsic factor essential for the absorption of vitamin B_{12} in the ileum; and (4) **argentaffin cells,** or **G cells,** which secrete **gastrin.** Gastrin is a hormone that stimulates parietal cells to secrete HCl and chief cells to secrete **pepsinogen,** which, with the action of HCl, becomes **pepsin,** a proteolytic enzyme.

Intestinal Tract The intestine is a long, highly convoluted hollow tubular structure that extends from the **pyloric** end of the stomach to the **anus.** The intestine is also composed of four identifiable layers: the **adventitia** (serosa, if the intestine is covered by the peritoneal fold of the mesentery), the **muscularis,** the **submucosa,** and the **mucosa.** The mucosal epithelium and the lamina propria below it form finger-like projections (**villi**), folds, or a flat surface.

There are two major subdivisions of the intestine, the **small intestine** and the **large intestine.** The 23-foot-long small intestine is further subdivided into the **duodenum,** the **jejunum,** and the **ileum.** The large intestine or **colon** is about five feet long, and its diameter is almost twice that of the small intestine. It is subdivided into the **cecum, appendix, colon (ascending, transverse, descending,** and **sigmoid),** and **rectum.** The mucosa of the colon lacks the villi that identify the small intestine. In the colon, mucus-secreting glands are tightly packed and consist of goblet cells mixed with columnar cells.

The muscularis, submucosa, and mucosa of the cecum fold twice at the juncture of the ileum and cecum to form the **ileocecal valve.** The vermiform appendix is a blind tubular structure that extends from the pouch-like structure of the **cecum. Peyer's patches** are solitary lymphoid nodules found in the lamina propria of the **colon** and **appendix** (also present in the small intestine). The outer smooth layer of the muscularis of the cecum and colon is modified to form three thick muscular bands or **taeniae coli.**

Rectum   The rectum is similar to the colon but lacks the taeniae coli bands of smooth muscle. The epithelium changes from columnar to nonkeratinized stratified squamous epithelium at the rectoanal junction.

Large **apocrine glands (cutaneous glands)** and hair surround the anal opening at the very distal end of the gastrointestinal tract. The submucosa in this region is highly vascularized and forms a plexus of **hemorrhoidal vessels.** The muscularis in the region is also modified to form the **internal anal sphincter.** Distally, the smooth muscle fibers combine with the striated muscle fibers to form the **external anal sphincter.**

Liver   The liver is the largest organ in the body and is located below the **diaphragm.** It is partially separated into four lobes: the **right, left, caudate,** and **quadrate lobes.** The right lobe is the largest. The lobes are divided into hexagonal lobules. Within the lobules are anastomosing **hepatic cords** composed of **hepatocytes** that radiate in all directions from the central vein. The hepatic sinusoids within the hepatic cords are lined with endothelial cells, reticulate fibers, and phagocytic **von Kupffer cells.**

Several lobules share a **portal area** that contains a branch of the **hepatic artery,** a branch of the **portal vein,** a small **bile duct,** and possibly a **lymphatic channel.** The liver **acinus** is a smaller lobule associated with the blood supply to the liver lobules. The terminal branches of the portal vein and hepatic artery supply blood to the acinus. Bile from the acinus is drained through the bile duct. Bile canaliculi are tiny channels between adjacent hepatocytes. The bile **canaliculi** collect bile from the hepatocytes, transport it through a network of canaliculi in the parenchyma of the lobules, and empty it into bile ducts in the portal area.

Gallbladder   The gallbladder is a blind sac located under the large (right) lobe of the liver. The short **cystic duct** connects the hepatic duct to the gallbladder. The mucosa of the cystic duct folds toward the lumen, forming **spiral valves** with smooth muscle in the body walls. The muscles prevent the cystic duct from collapsing during intense pressure. The epithelial lining of the mucosa is of simple columnar epithelium. The lamina propria, submucosa, muscularis, and serosa underlie the epithelium.

Bile produced in the liver is emptied into the hepatic ducts and forced backward into the gallbladder, where it is concentrated and stored.

Pancreas   The pancreas is retroperitoneal and is the second largest gland in the body. It functions as both an **endocrine gland** and an **exocrine gland.** The pancreas is divided into small lobules separated by delicate strands of connective and adipose tissue. The lobules contain pancreatic islets called the **islets of Langerhans,** which are small endocrine cell concentrations. These islets produce the hormones **glucagon, insulin, somatostatin, vasoactive intestinal peptide (VIP), substance P, motilin** (motilin is also secreted by the gut), and **pancreatic gastrin.**

The exocrine pancreas is a **tubuloacinar gland.** The secretory units are generally a combination of flask- and tube-shaped **acini.** The acinar cells secrete the digestive enzymes **lipase, amylase, trypsin, chymotrypsin, elastase, carboxypeptidases A** and **B, deoxyribonuclease, ribonuclease,** and **bicarbonate** to neutralize **chyme HCl.** The enzymes and bicarbonate are emptied into the duodenum of the gastrointestinal tract.

Oral Cavity

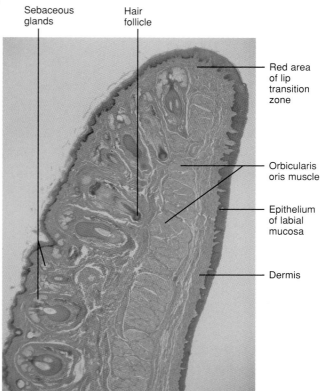

Sebaceous glands

Hair follicle

Red area of lip transition zone

Orbicularis oris muscle

Epithelium of labial mucosa

Dermis

FIGURE 15.1

Light micrograph (LM) of a sagittal section through a lip. Lip tissue is primarily orbicularis oris striated muscle fibers, fibroelastic connective tissue, hair follicles, sebaceous glands, and sweat glands. The outer surface of a lip is covered with a thin layer of stratified squamous epithelium that gradually blends with the mucous membrane that lines the oral cavity. At the transition zone, numerous connective tissue papillae of the underlying dermis form the line of demarcation between the outer and inner linings of the lips. (20×)

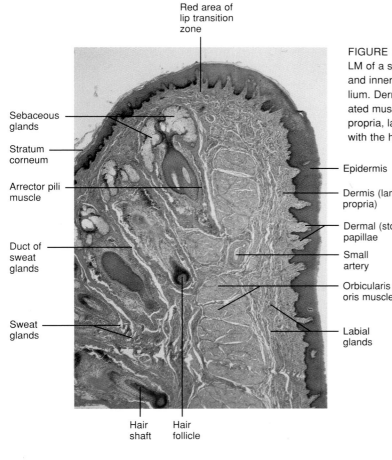

Red area of lip transition zone

Sebaceous glands

Stratum corneum

Arrector pili muscle

Duct of sweat glands

Sweat glands

Hair shaft

Hair follicle

Epidermis

Dermis (lamina propria)

Dermal (stomal) papillae

Small artery

Orbicularis oris muscle

Labial glands

FIGURE 15.2

LM of a sagittal section of the lip at a higher magnification. The outer and inner surfaces of the lip are lined with stratified squamous epithelium. Dermal (stomal) papillae, hair follicles, sebaceous glands, striated muscle fibers, sweat glands, a neurovascular bundle, lamina propria, labial mucosa, and arrector pili (smooth muscle associated with the hair follicle) can be identified below the epithelium. (40×)

Papillae   Epithelium

Lamina
propria

Mucous
glands

Muscle
(l.s.)

Muscle
(t.s.)

Small
artery

Lower
epithelium
and lamina
propria

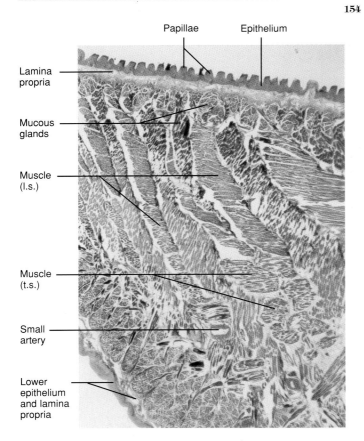

FIGURE 15.3
LM of a cross section through the tongue. The superior surface of
the tongue is predominantly keratinized stratified squamous
epithelium with elevated small structures known as papillae. The
inferior surface mucous layer is nonkeratinized stratified squamous epithelium. Below the epithelium are crisscrossing, interlacing bundles of skeletal muscles, dermal (stomal) papillae,
mucosa, serous and mucus acini of lingual glands, blood vessels,
nerves, connective tissue of the lamina propria, and excretory
ducts associated with the lingual glands. (20×)

FIGURE 15.4
LM of a section through filiform papillae. The papillae
are lined with keratinized epithelium. Below the
epithelium is the lamina propria of the mucosa,
skeletal muscle, and the lingual glands. (40×)

Filiform   Lamina
papillae   propria

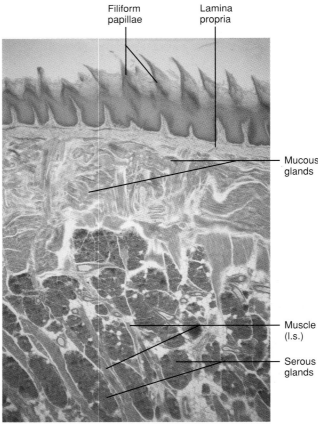

Mucous
glands

Muscle
(l.s.)

Serous
glands

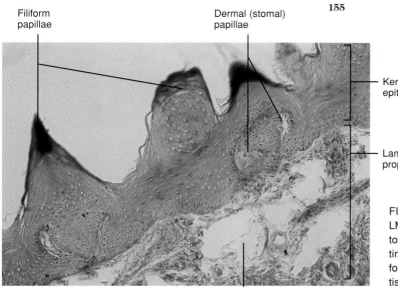

Filiform papillae

Dermal (stomal) papillae

Keratinized epithelium

Lamina propria

Adipose tissue

FIGURE 15.5
LM of a cross section through the upper surface of the tongue. The micrograph shows filiform papillae with keratinized epithelium, dermal (stomal) papillae, and mucosal folds of the lamina propria. Adipose tissue and connective tissue are also visible. (200×)

FIGURE 15.6
LM of a vallate or circumvallate papilla with a few fungiform papillae. Only the dorsal surface vallate papilla is keratinized. Taste buds are present on the lateral surface. Below the surface epithelium are the connective tissue skeletal muscle, mucous glands, ducts, adipose tissue, and secretory ducts associated with the glands. (4×)

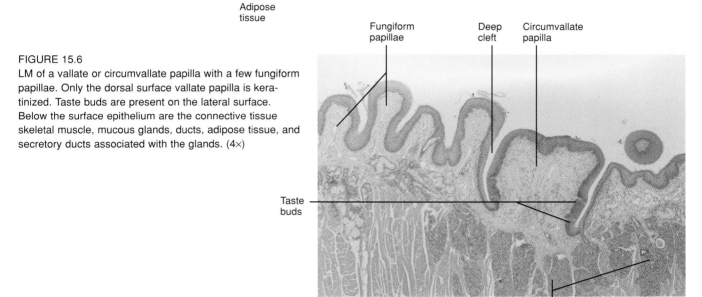

Fungiform papillae

Deep cleft

Circumvallate papilla

Taste buds

Von Ebner glands

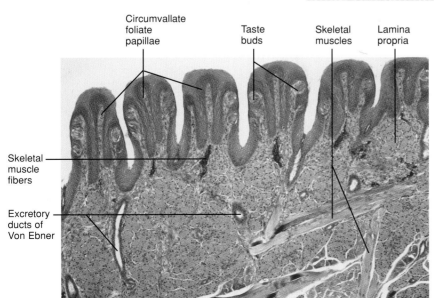

Circumvallate foliate papillae

Taste buds

Skeletal muscles

Lamina propria

Skeletal muscle fibers

Excretory ducts of Von Ebner

FIGURE 15.7
LM of a section through the foliate papillae. These papillae are folds of the mucosa near the base of the tongue. The taste buds are located on the lateral surfaces of the papillae. Also present are dermal papillae, mucosal glands, skeletal muscle, and several secretory ducts in section. Foliate papillae are not prominent in humans but are well developed in lower mammals. (100×)

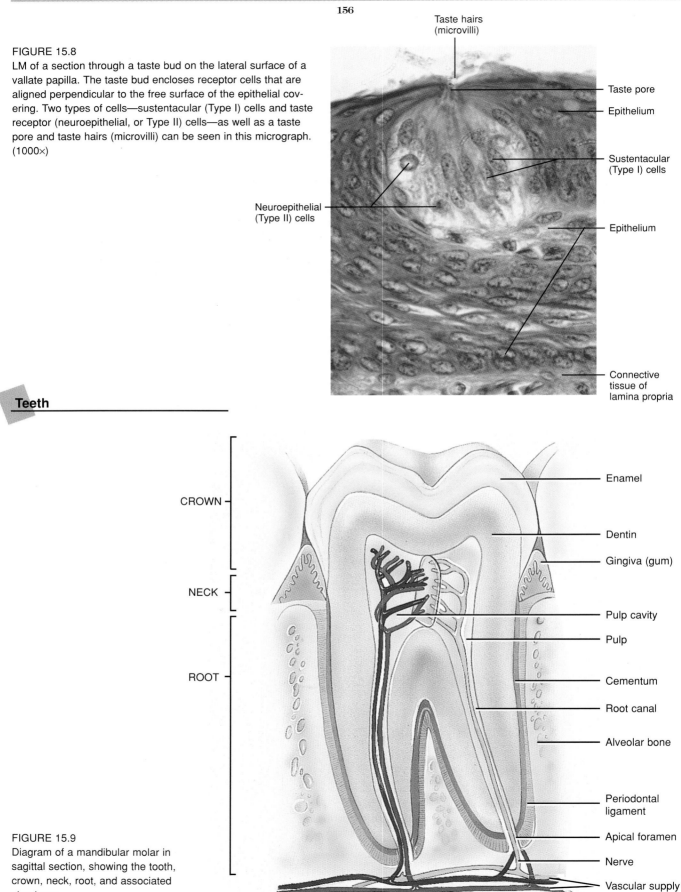

FIGURE 15.8
LM of a section through a taste bud on the lateral surface of a vallate papilla. The taste bud encloses receptor cells that are aligned perpendicular to the free surface of the epithelial covering. Two types of cells—sustentacular (Type I) cells and taste receptor (neuroepithelial, or Type II) cells—as well as a taste pore and taste hairs (microvilli) can be seen in this micrograph. (1000×)

Taste hairs (microvilli)

Taste pore

Epithelium

Sustentacular (Type I) cells

Neuroepithelial (Type II) cells

Epithelium

Connective tissue of lamina propria

Teeth

CROWN

NECK

ROOT

Enamel

Dentin

Gingiva (gum)

Pulp cavity

Pulp

Cementum

Root canal

Alveolar bone

Periodontal ligament

Apical foramen

Nerve

Vascular supply

FIGURE 15.9
Diagram of a mandibular molar in sagittal section, showing the tooth, crown, neck, root, and associated structures.

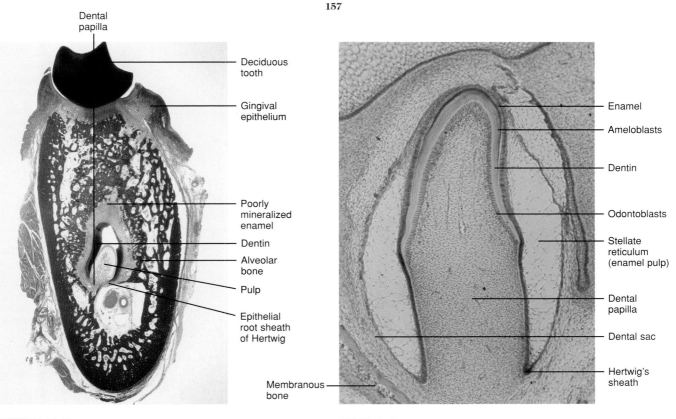

Dental papilla

Deciduous tooth

Gingival epithelium

Poorly mineralized enamel

Dentin

Alveolar bone

Pulp

Epithelial root sheath of Hertwig

Enamel

Ameloblasts

Dentin

Odontoblasts

Stellate reticulum (enamel pulp)

Dental papilla

Dental sac

Hertwig's sheath

Membranous bone

FIGURE 15.10
LM of a sagittal section through a deciduous tooth and a developing permanent tooth. The deciduous tooth is at the stage where it will be replaced by progressive development of the primordium of the permanent tooth. (1×)

FIGURE 15.11
LM of a sagittal section through the dental primordium, showing early stages in the deposition of dentin and enamel, and the formation of dental papilla. The tooth is embedded in the bony alveolus surrounded by the dental follicle. (100×)

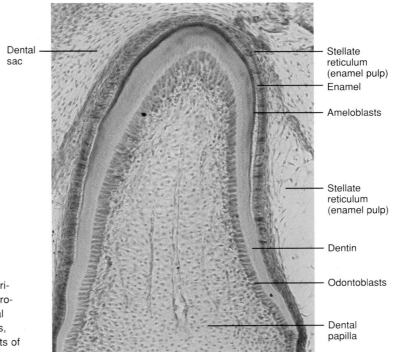

Dental sac

Stellate reticulum (enamel pulp)

Enamel

Ameloblasts

Stellate reticulum (enamel pulp)

Dentin

Odontoblasts

Dental papilla

FIGURE 15.12
LM of a sagittal section through the dental primordium, at a higher magnification. The micrograph shows detailed differentiation of dental papilla, dentin, enamel and surrounding cells, odontoblasts, and ameloblasts. The remnants of the dental lamina are also visible. (200×)

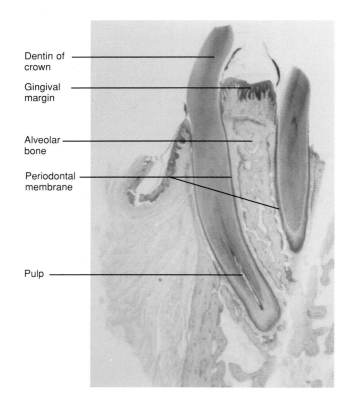

Dentin of crown

Gingival margin

Alveolar bone

Periodontal membrane

Pulp

FIGURE 15.13
LM of a sagittal section through two newly erupted teeth extending beyond the gingiva. The roots of the teeth are embedded in the bony alveolus or osseous socket. Enamel is absent as a result of its removal through decalcification prior to infiltration and sectioning. (20×)

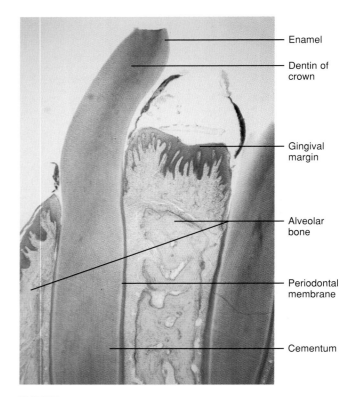

Enamel

Dentin of crown

Gingival margin

Alveolar bone

Periodontal membrane

Cementum

FIGURE 15.14
LM of a newly erupted tooth in sagittal section. The tooth is surrounded by keratinized squamous epithelium forming the gingiva. Connective tissue and the periodontal membrane and dermal papillae associated with the dermis can be seen below the epidermis. (40×)

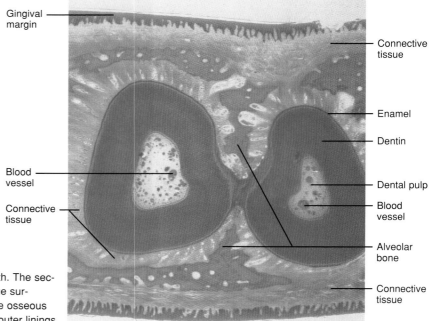

Gingival margin

Connective tissue

Enamel

Dentin

Dental pulp

Blood vessel

Blood vessel

Alveolar bone

Connective tissue

Connective tissue

FIGURE 15.15
LM of a cross section through the roots of two teeth. The section shows the pulp cavity, dentin, connective tissue surrounding the root, and cancellous bone forming the osseous socket. The epithelial gingiva forms the inner and outer linings of the gum. (20×)

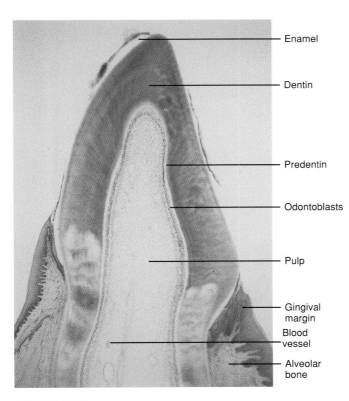

FIGURE 15.16
LM of a sagittal section through the anatomical crown and part of the anatomical root of a tooth. The enamel of the crown is poorly preserved in the section. However, the dentin, pulp cavity, gingiva, and surrounding connective tissue can be identified. (20×)

Enamel

Dentin

Predentin

Odontoblasts

Pulp

Gingival margin

Blood vessel

Alveolar bone

Salivary Glands

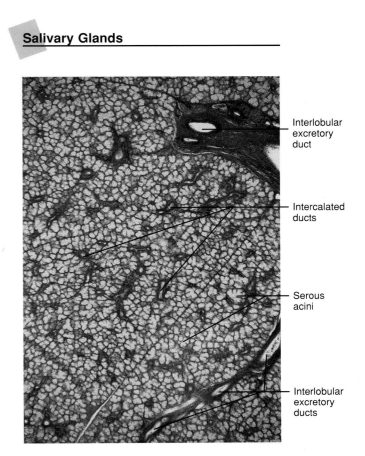

Interlobular excretory duct

Intercalated ducts

Serous acini

Interlobular excretory ducts

FIGURE 15.17
LM of a cross section through a parotid salivary gland. Classified as a compound serous tubuloacinar gland, the parotid is surrounded by a connective tissue capsule. The parotid gland is divided into lobes and lobules separated by connective tissue septa. The secretory structures that facilitate the excretory process consist of serous acini. The acini are surrounded by pyramid-shaped myoepithelial cells. The secretory products from the acini are emptied into narrow channels, the intercalated ducts, which join the larger striated ducts. The striated ducts then empty into interlobular excretory ducts. Finally, the interlobular ducts drain into larger lobar ducts. The lobular ducts join to form the Stensen's ducts of the parotid gland. (40×)

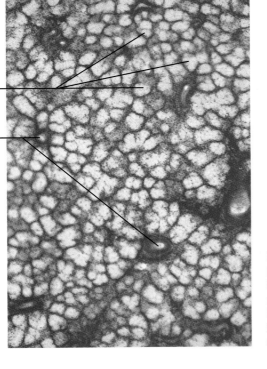

Serous acini

Intercalated ducts

FIGURE 15.18
LM of a cross section through a parotid gland lobule, at a higher magnification. This gland is characterized by large numbers of excretory ducts and adipose cells. At this higher magnification, small secretory granules are visible in the cells. Contractile myoepithelial cells surround the serous acini. (100×)

FIGURE 15.19
LM of a cross section of a sublingual mixed tubuloacinar
gland. The gland is composed of both mucus and serous
acini, with mucus acini predominating. Adipose tissue and
myoepithelial cells that surround the acini are present in
the gland. Intercalated excretory ducts are infrequent.
Interlobular excretory ducts prevail throughout the lobules
of the sublingual glands. Connective tissue septa separate the lobules. (40×)

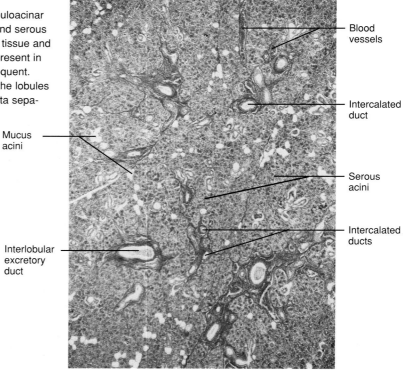

Blood
vessels

Intercalated
duct

Mucus
acini

Serous
acini

Intercalated
ducts

Interlobular
excretory
duct

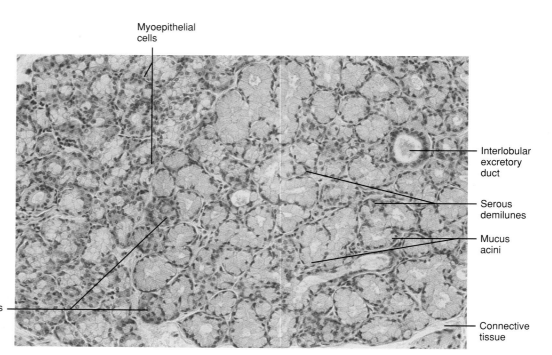

Myoepithelial
cells

Interlobular
excretory
duct

Serous
demilunes

Mucus
acini

Connective
tissue

Serous
acini

FIGURE 15.20
LM of sublingual gland at a higher magnification. Both serous and mucus acini are present,
with mucus acini predominating in the tissue. Visible in the micrograph are nonstriated intercalated excretory ducts, clear serous demilunes, and mucus acini. (200×)

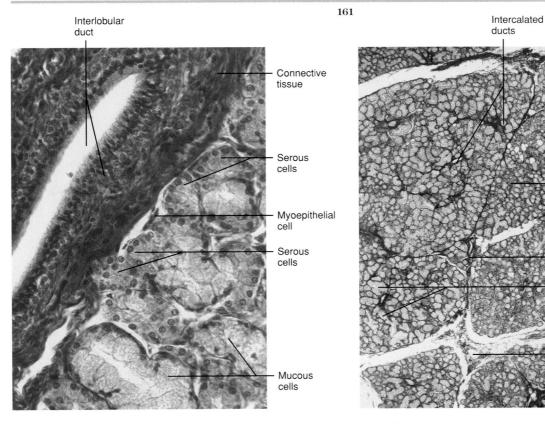

Interlobular duct

Connective tissue

Serous cells

Myoepithelial cell

Serous cells

Mucous cells

Intercalated ducts

Serous acini

Interlobular septa

Mucus acini

Interlobular excretory duct

Blood vessel

FIGURE 15.21

LM of a sublingual gland, exhibiting a large interlobular excretory duct surrounded by collagen fibers, mucus acini, and serous demilunes. The interlobular (striated) excretory ducts are lined with cuboidal cells. The basal cytoplasm of the cuboidal cells forms deep enfoldings, resulting in a striated appearance. (400×)

FIGURE 15.22

LM of a submandibular gland in cross section. The submandibular gland is a compound tuboalveolar gland surrounded by a fibrous capsule with septa, lobes, and lobules. Also present are myoepithelial cells and a well-established duct system with fewer ducts than the parotid gland. The majority of the secretory units in the gland are serous. The micrograph shows the connective tissue septa, lobules, interlobular ducts, and secondary ducts. (40×).

FIGURE 15.23

LM of a cross section through a submandibular gland, at a higher magnification. The micrograph shows the submandibular gland as a mixed gland in which mucous cells are surrounded by serous cells. The serous demilunes of von Ebner, crescent-shaped encroachments of serous cells into mucous cells, can also be identified. (400×)

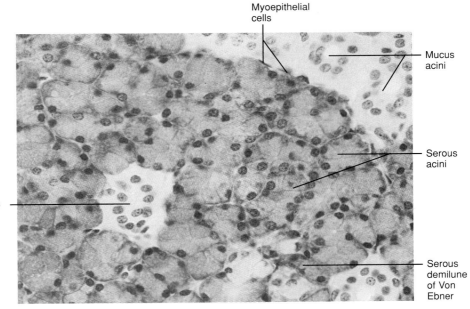

Myoepithelial cells

Mucus acini

Serous acini

Serous demilune of Von Ebner

Mucus acini

The Gastrointestinal Tract

Esophagus   Stomach   Small intestine   Large intestine

Duct of gland outside tract
(such as salivary gland or
pancreas)

Longitudinal muscle

Superficial

Circular muscle

Gland in submucosa

Muscularis mucosae

Myenteric
plexus
(plexus of
Auerbach)

Lamina propria

Epithelium

Deep

Villus

Lymphatic
nodule

Lumen

Submucosal plexus
(plexus of Meissner)

MUCOSA

SUBMUCOSA

Gland in submucosa

Glands in mucosa

MUSCULARIS

Mesentry

SEROSA

DANK

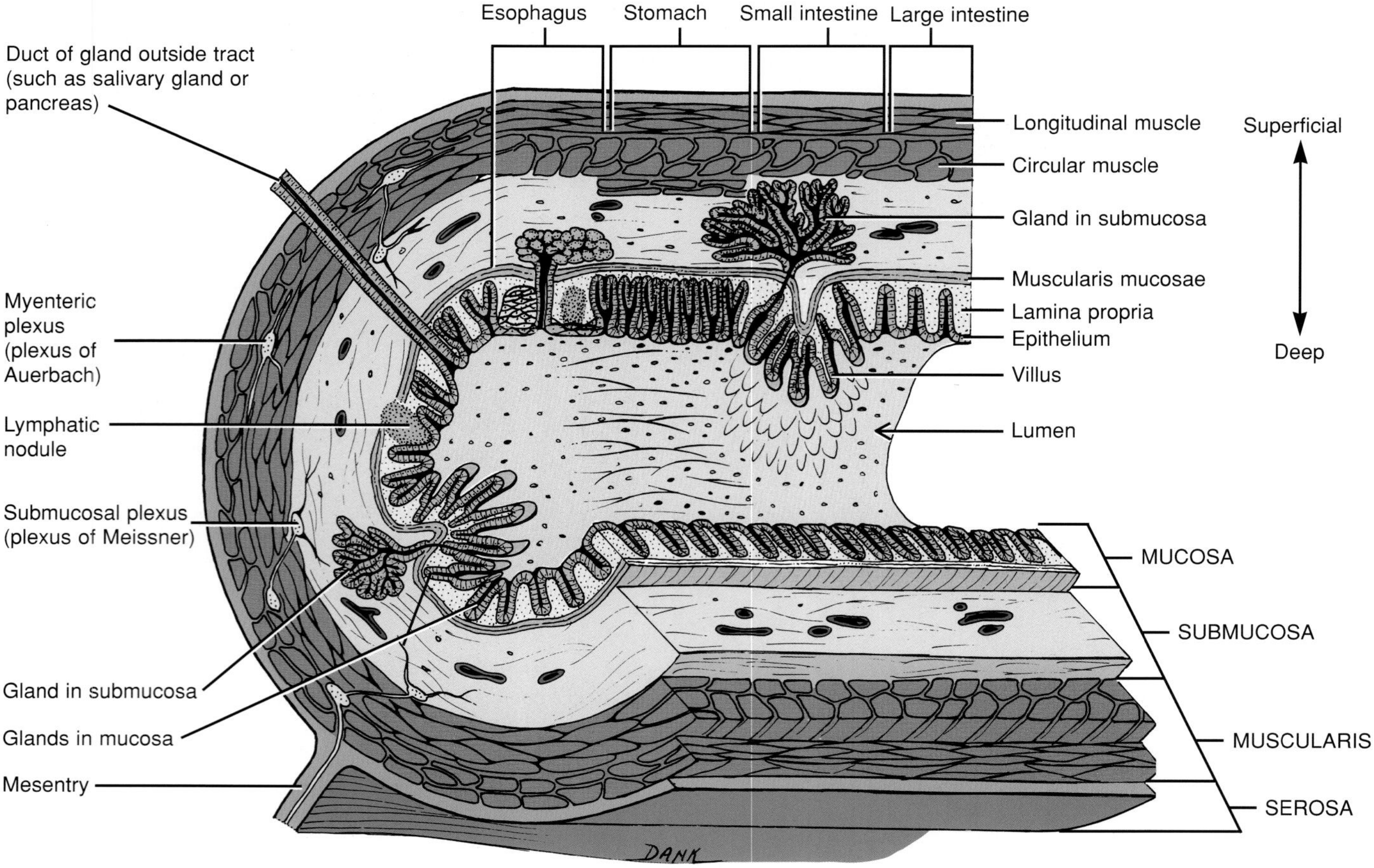

FIGURE 15.24
Sectional diagram of the gastrointestinal tract through the regions of
the esophagus, stomach, small intestine, and large intestine. Also
shown are subdivisions of the body wall and related structures.

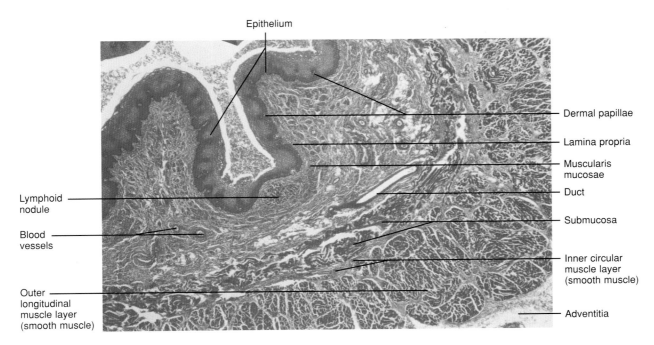

FIGURE 15.25
LM of a complete cross section of an esophagus. The section illustrates the typical body wall structure maintained throughout the remainder of the gastrointestinal tract. An exception is the mucosa, which is modified according to function. In the esophagus, the mucosa consists of stratified squamous epithelium (nonkeratinized), loosely arranged lamina propria, and muscularis mucosae. The submucosa consists of the submucosal plexus (plexus of Meissner). The muscularis is subdivided into inner circular and outer longitudinal layers of smooth muscle with autonomic myenteric plexus between the muscle layers. The adventitia forms the outermost covering of the alimentary canal. (1×)

FIGURE 15.26
LM of a cross section through the wall of the esophagus. The surface epithelium of the mucosa is nonkeratinized stratified squamous. The lamina propria is narrow with an outer band of muscularis mucosae. Also visible in the micrograph are a lymphoid nodule, submucosa, muscularis, and the adventitia. (20×)

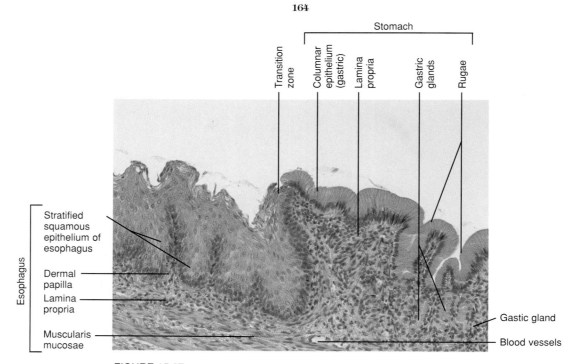

Stomach

Transition zone

Columnar epithelium (gastric)

Lamina propria

Gastric glands

Rugae

Esophagus

Stratified squamous epithelium of esophagus

Dermal papilla

Lamina propria

Muscularis mucosae

Gastic gland

Blood vessels

FIGURE 15.27
LM of a longitudinal section through the esophagus-stomach junction. There is an abrupt change in the epithelium from the stratified nonkeratinized squamous epithelium of the esophagus to the simple columnar epithelium (with gastric pits) of the stomach. (100×)

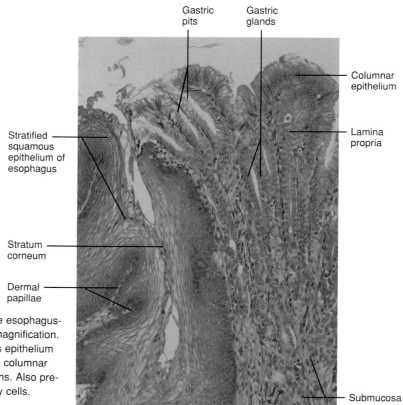

Gastric pits

Gastric glands

Columnar epithelium

Lamina propria

Stratified squamous epithelium of esophagus

Stratum corneum

Dermal papillae

Submucosa

FIGURE 15.28
LM of a longitudinal section through the esophagus-cardiac-stomach junction, at a higher magnification. The nonkeratinized stratified squamous epithelium of the esophagus terminates where the columnar epithelium of the cardiac-stomach begins. Also present are gastric pits lined with secretory cells. (200×)

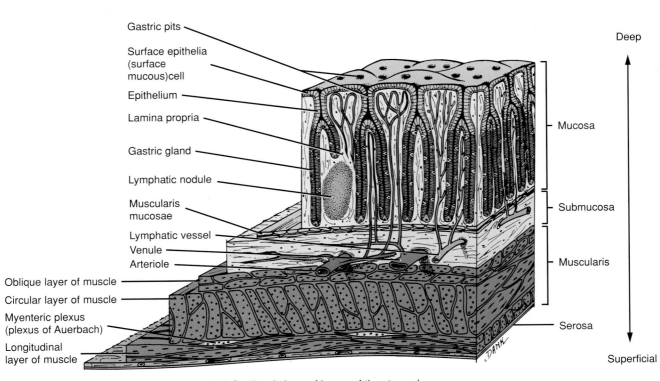

Gastric pits

Surface epithelia (surface mucous)cell

Epithelium

Lamina propria

Gastric gland

Lymphatic nodule

Muscularis mucosae

Lymphatic vessel

Venule

Arteriole

Oblique layer of muscle

Circular layer of muscle

Myenteric plexus (plexus of Auerbach)

Longitudinal layer of muscle

Deep

Mucosa

Submucosa

Muscularis

Serosa

Superficial

(a) Sectional views of layers of the stomach

FIGURE 15.29
(a) Diagram of the layers of the stomach body wall and important structures. The stomach body wall is divided into the serosa, muscularis, submucosa, and mucosa. The mucosa forms folds called rugae, and the epithelium is comprised of columnar cells.
(b) Diagram of the mucosa, displaying glands and gastric pits present in the stomach body wall.

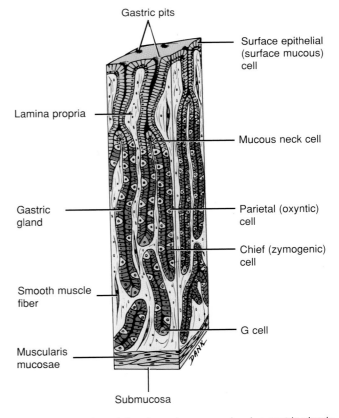

Gastric pits

Surface epithelial (surface mucous) cell

Lamina propria

Mucous neck cell

Gastric gland

Parietal (oxyntic) cell

Chief (zymogenic) cell

Smooth muscle fiber

G cell

Muscularis mucosae

Submucosa

(b) Sectional view of the stomach mucosa showing gastric glands

FIGURE 15.30
LM of a longitudinal section of the body wall of the stomach in a nondistended state. The micrograph illustrates the different regions of the body wall: the epithelium, lamina propria, and muscularis mucosae of the mucosa; the submucosa; the muscle layers of the muscularis; and the outermost covering, the adventitia. (40×)

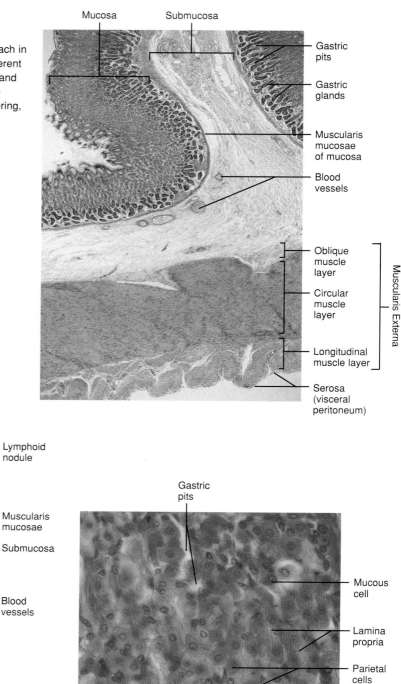

FIGURE 15.31
LM of a section through the body wall of the gastric fundus. The secretory tubules of the mucosa are closely spaced, and the gastric pits are shallow. The absence of goblet cells differentiates the gastric fundus from the colic (colon) mucosa, where goblet cells are in abundance. Also shown in the micrograph is a lymphoid nodule. (100×)

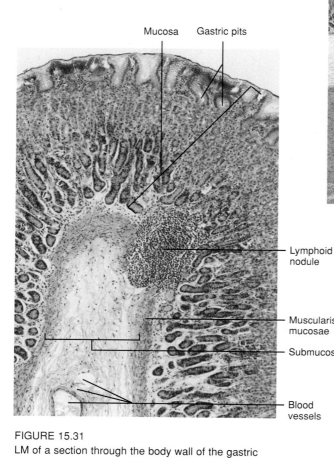

FIGURE 15.32
LM of a longitudinal section through the mucosal layer of the gastric fundus, at a higher magnification. Visible in the micrograph are gastric glands and their cells, chief cells (peptic or zymogenic cells), parietal cells (oxyntic cells), and mucoid neck cells. Also present are gastric pits and the lamina propria of the mucosa. (400×)

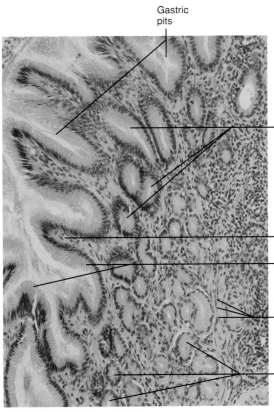

Gastric pits

Pyloric glands

Lamina propria

Epithelium (mucous and columnar cells)

Muscle fibers

Pyloric gland

FIGURE 15.33

LM of a section through the pyloric stomach. Visible in the micrograph are the mucosa with pyloric glands and the gastric pits. The pyloric gastric pits are much deeper than the gastric pits seen in the fundic or cardiac regions. The bases of the pits are coiled and are represented by cross sections through the glands. The surface epithelium is simple columnar as in other parts of the stomach. (200×)

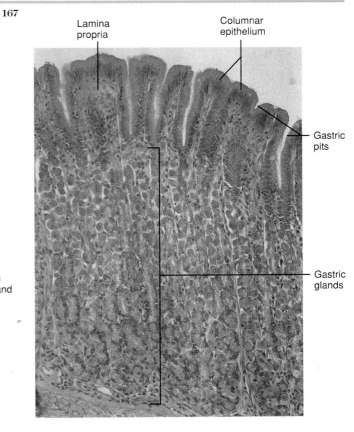

Lamina propria

Columnar epithelium

Gastric pits

Gastric glands

FIGURE 15.34

LM of the lining of the body wall of the stomach, illustrating mucosal gastric pits and the tall single layer of mucus-secreting columnar cells. Parietal and chief (peptic) cells are generally located at the bases of the gastric glands. (200×)

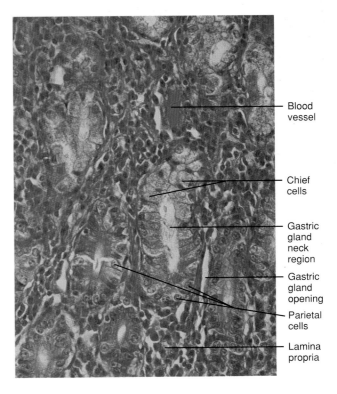

Blood vessel

Chief cells

Gastric gland neck region

Gastric gland opening

Parietal cells

Lamina propria

FIGURE 15.35

LM of a section at the base of a gastric gland. Chief cells (peptic cells) are the primary cell type in the lower third of the gastric glands. However, parietal cells are also interspersed with the chief cells at this level. The peptic cells can be identified by the presence of granular cytoplasm and basically placed nuclei. The parietal cells show centrally placed nuclei and an eosinophilic cytoplasm. (400×)

Small Intestine

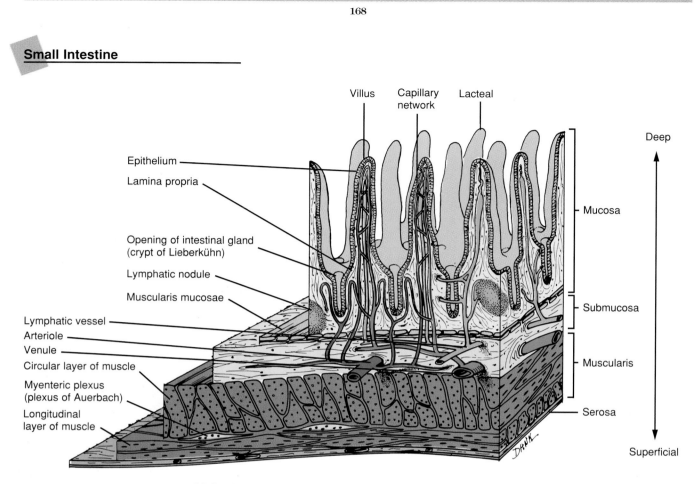

Villus   Capillary network   Lacteal

Deep

Epithelium

Lamina propria

Mucosa

Opening of intestinal gland
(crypt of Lieberkühn)

Lymphatic nodule

Muscularis mucosae

Submucosa

Lymphatic vessel

Arteriole

Venule

Circular layer of muscle

Myenteric plexus
(plexus of Auerbach)

Muscularis

Longitudinal
layer of muscle

Serosa

Superficial

(a) Sectional views of layers of the small intestine showing villi

FIGURE 15.36
(a) Diagram of the small intestine and enlarged villus. Identified in
cross section are the layers of tissue: serosa, muscularis (note the
absence of oblique muscle fibers), submucosa, and mucosa. The
mucosa has modified to form villi, the finger-like projections.
(b) Diagram of the mucosal villus, submucosa, muscularis, and
serosa.

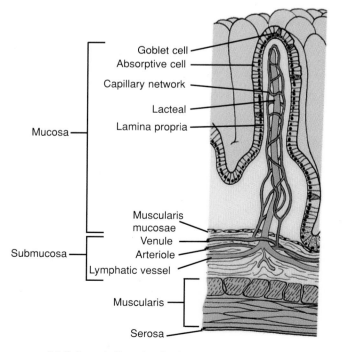

Goblet cell

Absorptive cell

Capillary network

Lacteal

Lamina propria

Mucosa

Muscularis
mucosae

Venule

Submucosa

Arteriole

Lymphatic vessel

Muscularis

Serosa

(b) Enlarged villus showing lacteal and capillaries

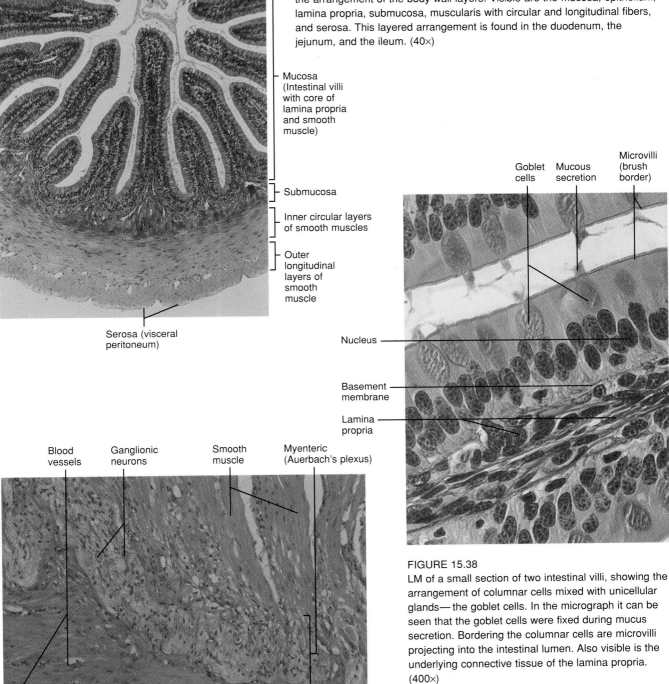

Intestinal lumen

FIGURE 15.37

LM of a cross section through the body wall of the small intestine, showing the arrangement of the body wall layers. Visible are the mucosa, epithelium, lamina propria, submucosa, muscularis with circular and longitudinal fibers, and serosa. This layered arrangement is found in the duodenum, the jejunum, and the ileum. (40×)

Mucosa (Intestinal villi with core of lamina propria and smooth muscle)

Submucosa

Inner circular layers of smooth muscles

Outer longitudinal layers of smooth muscle

Serosa (visceral peritoneum)

Goblet cells   Mucous secretion   Microvilli (brush border)

Nucleus

Basement membrane

Lamina propria

FIGURE 15.38

LM of a small section of two intestinal villi, showing the arrangement of columnar cells mixed with unicellular glands—the goblet cells. In the micrograph it can be seen that the goblet cells were fixed during mucus secretion. Bordering the columnar cells are microvilli projecting into the intestinal lumen. Also visible is the underlying connective tissue of the lamina propria. (400×)

Blood vessels   Ganglionic neurons   Smooth muscle   Myenteric (Auerbach's plexus)

Smooth muscle

FIGURE 15.39

LM of myenteric (Auerbach's) plexus located between the circular and longitudinal muscular layers of the muscularis. Parasympathetic preganglionic fibers innervate the smooth muscle cells, resulting in the stimulation and augmentation of muscle tone and peristaltic activity of the gastrointestinal tract. (200×)

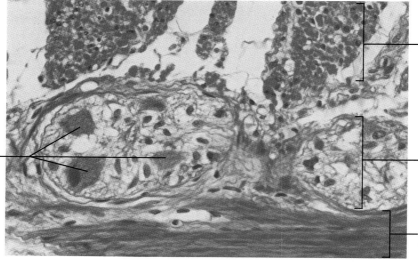

Longitudinal smooth muscle

Ganglionic neurons

Myenteric (Auerbach's) plexus of nerve cells and fibers

Circular smooth muscle

FIGURE 15.40
LM of a section through a myenteric (Auerbach's) plexus, at a higher magnification. The plexus is located at the junction of the circular and longitudinal smooth muscle fibers of the muscularis. Neurons can be seen within the plexus. (400×)

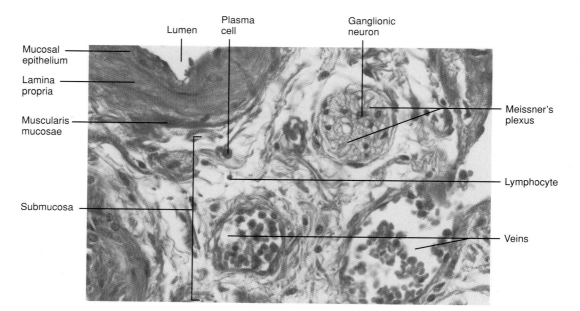

Lumen

Plasma cell

Ganglionic neuron

Mucosal epithelium

Lamina propria

Muscularis mucosae

Submucosa

Meissner's plexus

Lymphocyte

Veins

FIGURE 15.41
LM of a cross section through the submucosa of the gastrointestinal tract. Small clusters of parasympathetic ganglia (Meissner's plexus) can be seen in the connective tissue of the submucosa. The plexus is close to two blood vessels and the muscularis mucosae of the mucosa. (400×)

Villi

Mucosa

Submucosa

Brunner's glands

Circular muscle layer

Longitudinal muscle layer

FIGURE 15.42
LM of a cross section through the body wall of the duodenum. Visible in the micrograph are the mucosa with finger-like villi, the lamina propria, a thin layer of muscularis mucosae, the submucosa with blood vessels and Brunner's glands, the muscularis with inner circular and outer longitudinal smooth muscle cells, and a narrow covering of the serosa. (40×)

FIGURE 15.43
LM of a cross section through the duodenum, at a higher magnification. Visible are mucosal villi, lamina propria, and submucosa with connective tissue, blood vessels, and Brunner's glands. A small portion of the circular layer of smooth muscle cells of the muscularis can also be seen. (100×)

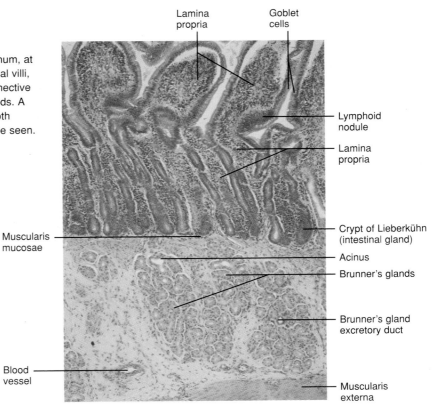

Lamina propria

Goblet cells

Lymphoid nodule

Lamina propria

Crypt of Lieberkühn (intestinal gland)

Acinus

Brunner's glands

Brunner's gland excretory duct

Muscularis mucosae

Blood vessel

Muscularis externa

FIGURE 15.44
LM of Brunner's glands as seen in the submucosa of the duodenum. The Brunner's
glands are compound, tubular, mucus-
secreting glands. Their secretory ducts open
into the intestinal crypts of the mucosa. (200×)

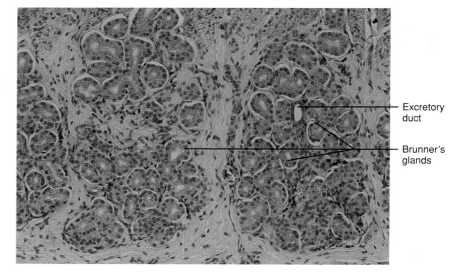

— Excretory
duct

— Brunner's
glands

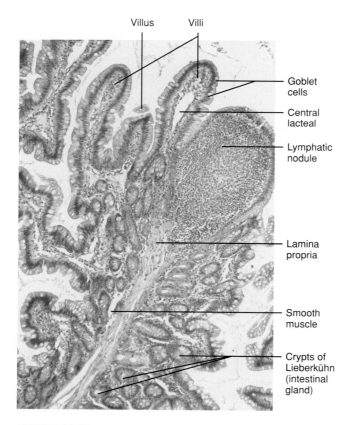

Villus   Villi

Goblet
cells

Central
lacteal

Lymphatic
nodule

Lamina
propria

Smooth
muscle

Crypts of
Lieberkühn
(intestinal
gland)

FIGURE 15.45
LM of a cross section through the mucosa of the
duodenum. Seen in the micrograph are many villi in
cross and longitudinal sections, a large lymphoid
nodule, lamina propria, and simple columnar epithelium cells mixed with goblet cells bordering the villi.
(100×)

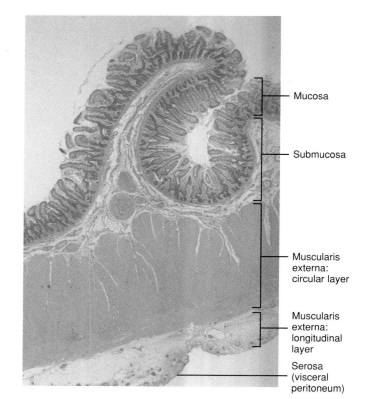

Mucosa

Submucosa

Muscularis
externa:
circular layer

Muscularis
externa:
longitudinal
layer

Serosa
(visceral
peritoneum)

FIGURE 15.46
LM of a cross section through the jejunum, showing
the villi that line the folds (the plica circulares) of the
small intestine. The plica circulares constitutes both
the mucosa and the submucosa. The villi display
columnar epithelium lining toward the lumen. Also
present are the muscularis and the outer covering,
the serosa. (20×)

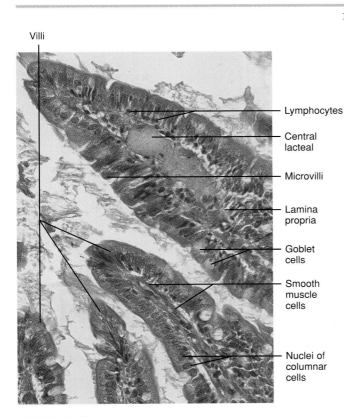

Villi

Lymphocytes

Central lacteal

Microvilli

Lamina propria

Goblet cells

Smooth muscle cells

Nuclei of columnar cells

FIGURE 15.47
LM of jejunal villi at a higher magnification. Columnar cells mixed with goblet cells form the epithelial lining of the villi. In the lamina propria, elongated myoepithelial cells, lacteals, and blood vessels are visible. (400×)

Villi   Crypts of Lieberkühn (intestinal glands)

Lymphatic nodule reaching surface of mucosa

Lamina propria

Peyer's patches (lymphatic nodules)

Muscularis mucosae

Submucosa

Muscularis externa: circular layer

Muscularis externa: longitudinal layer

FIGURE 15.48
LM of a cross section through the ileum. The micrograph shows the lumen of the intestinal tract, villi sectioned at different planes, crypts of Lieberkühn, lamina propria, Peyer's patches (lymphoid nodules), muscularis mucosae, submucosa, muscularis externa, and serosa. In the ileum, the Peyer's patches form aggregates spaced at different intervals. (40×)

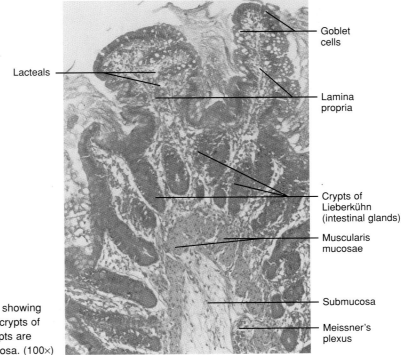

Lacteals

Goblet cells

Lamina propria

Crypts of Lieberkühn (intestinal glands)

Muscularis mucosae

Submucosa

Meissner's plexus

FIGURE 15.49
LM of a cross section through the ileum showing the intestinal villi, intestinal glands, and crypts of Lieberkühn at the base the villi. The crypts are located in the lamina propria of the mucosa. (100×)

FIGURE 15.50
LM of ileum villi in cross section, displaying microvilli.
The microvilli are finger-like projections extending from
the epithelial lining of the mucosa. The microvilli can
increase the absorptive surface of a cell to 30 times its
surface area. (1000×)

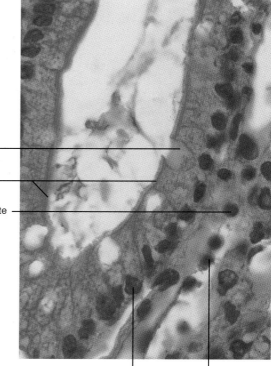

Goblet cell
Microvilli
Lymphocyte
Nucleus
Plasma cell

Submucosa — Paneth cells — Goblet cells — Crypts of Lieberkühn

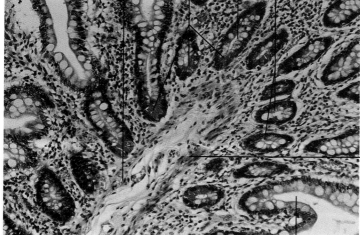

Muscularis mucosae

Crypt of Lieberkühn

FIGURE 15.51
LM of a cross section through the mucosal layer of
the ileum. Visible are small groups of pyramid-
shaped Paneth cells (located at the bases of crypts
of Lieberkühn) with conspicuous secretory granules.
Paneth cells secrete the bactericidal enzyme
lysozyme. Also visible in the micrograph are
columnar cells at the bases of the crypts, lamina
propria of the mucosa, and mucus-secreting goblet
epithelial cells. (200×)

Blood vessel — Paneth cells — Goblet cells — Lymphocytes

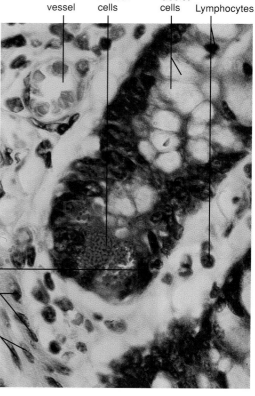

Epithelial cells
Lamina propria (connective tissue)
Fibroblasts

FIGURE 15.52
LM of Paneth and supporting cells of the crypt at a higher magnification. Pyramid-
shaped, clustered, lysozyme-secreting Paneth cells, goblet cells, crypt base
columnar cells, and surrounding connective tissue are visible. (1000×)

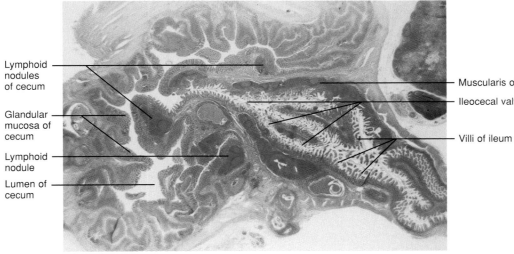

Lymphoid nodules of cecum

Glandular mucosa of cecum

Lymphoid nodule

Lumen of cecum

Muscularis of ileum

Ileocecal valve

Villi of ileum

FIGURE 15.53
LM of an ileocecal junction and a cone-shaped ileocecal valve in longitudinal section. The ileocecal valve is formed by the inward folding of the ileum mucosa, the submucosa, and the circular smooth muscle layer of the muscularis. The ileocecal valve opens into a larger distended cavity of the cecum, which can be seen in the micrograph as numerous folds of the cecal mucosa and submucosa. The epithelium of the ileum and the cecum in the region is columnar with goblet cells. Lymphoid tissue is prominent in the mucosal layer and forms solitary lymphatic nodules. (1×)

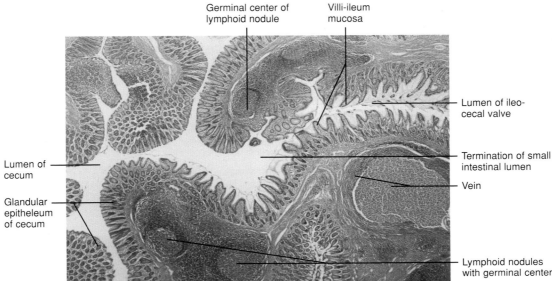

Germinal center of lymphoid nodule

Villi-ileum mucosa

Lumen of cecum

Glandular epitheleum of cecum

Lumen of ileo-cecal valve

Termination of small intestinal lumen

Vein

Lymphoid nodules with germinal centers

FIGURE 15.54
LM of a longitudinal section of the tip of the ileocecal valve as it opens into the cecum. The villi of the ileum are much shorter, the lamina propria is reduced, and most of the mucosa is occupied by lymphoid tissue. The folds of the cecum fill most of the cecal lumen, and the mucosal layer of the cecum lacks villi. (20×)

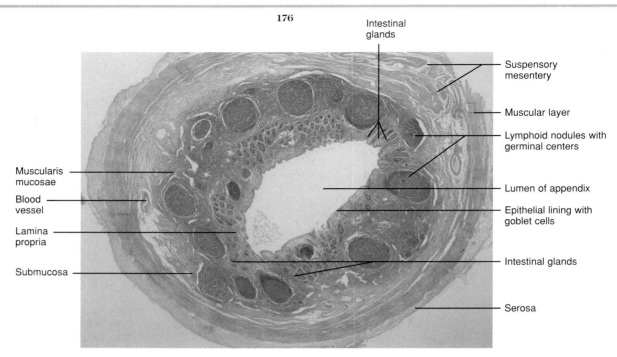

Intestinal glands

Suspensory mesentery

Muscular layer

Lymphoid nodules with germinal centers

Lumen of appendix

Epithelial lining with goblet cells

Intestinal glands

Serosa

Muscularis mucosae

Blood vessel

Lamina propria

Submucosa

FIGURE 15.55
LM of a cross section through the vermiform appendix at low magnification. The mucosa is lined with columnar epithelial cells mixed with goblet cells. The lamina propria shows large numbers of lymphatic nodules. Some nodules have germinal centers. Also present are the muscularis mucosae beneath the lamina propria, intestinal glands, diffused lymphoid tissue, submucosa, muscularis circular and longitudinal smooth muscle fibers, and serosa. (20×)

Large Intestine

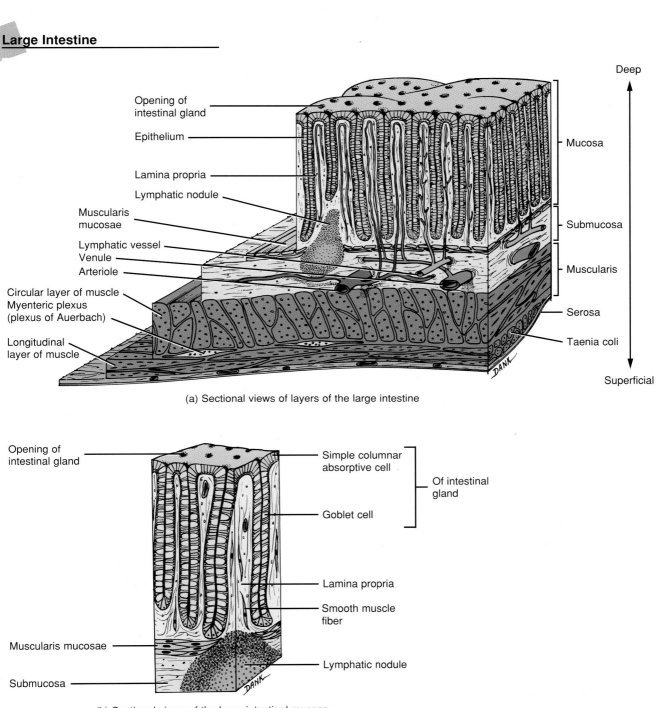

Deep

Opening of intestinal gland

Epithelium

Lamina propria

Lymphatic nodule

Muscularis mucosae

Lymphatic vessel

Venule

Arteriole

Circular layer of muscle
Myenteric plexus
(plexus of Auerbach)

Longitudinal
layer of muscle

Mucosa

Submucosa

Muscularis

Serosa

Taenia coli

Superficial

(a) Sectional views of layers of the large intestine

Opening of
intestinal gland

Simple columnar
absorptive cell

Of intestinal
gland

Goblet cell

Lamina propria

Smooth muscle
fiber

Muscularis mucosae

Lymphatic nodule

Submucosa

(b) Sectional views of the large intestinal mucosa

FIGURE 15.56
(a) Diagram of a cross section through the body wall of the large intestine (colon). There is a complete absence of villi. The inner surface is smooth and lined with columnar cells mixed with mucus-secreting goblet cells. (b) Diagram of the colon mucosa. A lymphoid nodule and the muscularis mucosae are represented in the lower portion of the diagram.

FIGURE 15.57
LM of a section through the large intestine (colon) at low magnification. Villi are absent, and the mucosal epithelium is rich with mucus-secreting goblet cells. Also, the lamina propria, crypts of Lieberkühn, muscularis mucosae, submucosa, and outer muscularis can be identified in the micrograph. The outer longitudinal muscle layer of the muscularis forms the taeniae coli muscle bands. (40×)

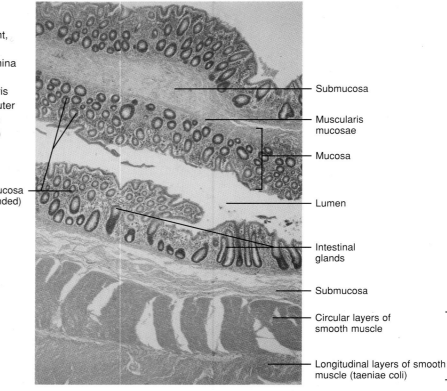

Folded mucosa (nondistended)

Submucosa

Muscularis mucosae

Mucosa

Lumen

Intestinal glands

Submucosa

Circular layers of smooth muscle

Longitudinal layers of smooth muscle (taeniae coli)

Muscularis

FIGURE 15.58
LM of a cross section through the colon, at a higher magnification. As seen in the micrograph, the colonic mucosal epithelium is lined with mucus-secreting goblet cells interspersed with the absorbing columnar cells. The lamina propria, muscularis mucosae, and submucosa form distinct layers. The crypts of Lieberkühn at the base of the mucosa and the outer muscularis can be identified. The outer longitudinal muscle layer of the muscularis forms the taeniae coli muscle bands. (200×)

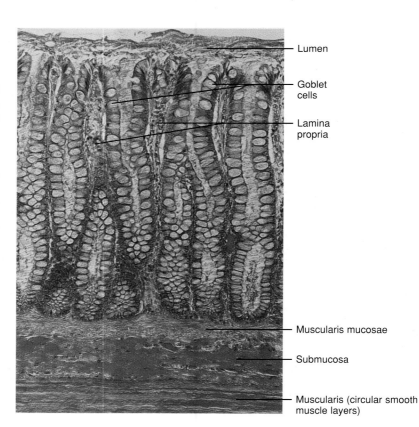

Lumen

Goblet cells

Lamina propria

Muscularis mucosae

Submucosa

Muscularis (circular smooth muscle layers)

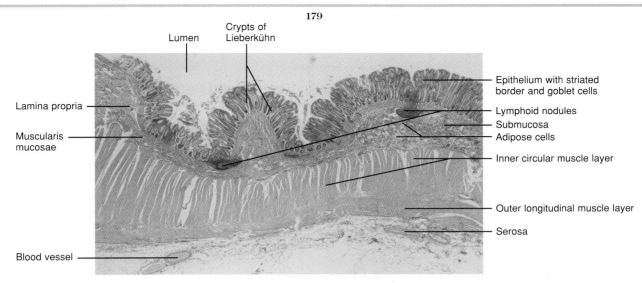

Lumen

Crypts of Lieberkühn

Lamina propria

Muscularis mucosae

Blood vessel

Epithelium with striated border and goblet cells

Lymphoid nodules

Submucosa

Adipose cells

Inner circular muscle layer

Outer longitudinal muscle layer

Serosa

FIGURE 15.59

LM of a cross section through the body wall of the rectum. The morphology of the rectum is similar to that of the colon. The mucosa is lined with goblet cells and tall columnar epithelial cells. The crypts of Lieberkühn, lamina propria, muscularis mucosae, and submucosa can be seen in the micrograph. The mucosa and submucosa show temporary folds. The taeniae coli muscle bands are absent in the rectum. The muscularis again forms into inner and outer layers of smooth muscle. (20×)

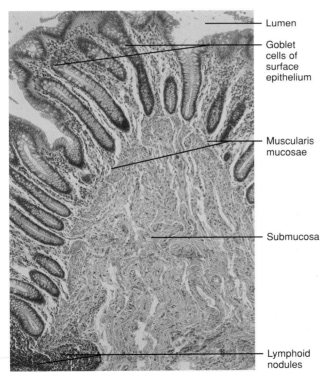

Lumen

Goblet cells of surface epithelium

Muscularis mucosae

Submucosa

Lymphoid nodules

Mucus

Goblet cells

Brush border

Columnar epithelial cells

Lymphocytes

Goblet cells

Lamina propria

Myoepithelial cells

FIGURE 15.60

LM of a cross section of rectal folds at a higher magnification. The intestinal glands extend much deeper into the mucosa than do the glands in the small intestine. The glands are essentially mucus-secreting goblet cells. Lamina propria connective tissue surrounds the glands. The submucosa is prominent, and lymphoid nodules can be identified at lower left. The muscularis mucosae is difficult to discern at this magnification. (100×)

FIGURE 15.61

LM of a small portion of the rectal mucosa. Simple columnar cells with brush border compose the epithelial lining. The tubular straight intestinal glands are primarily mucus-secreting goblet cells surrounded by the connective tissue of the lamina propria. Myoepithelial cells are present in the lamina propria. Mucus secreted by the epithelial cells can be seen at upper left. (400×)

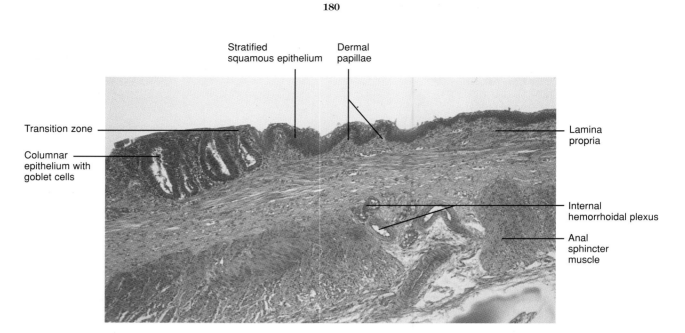

Stratified
squamous epithelium

Dermal
papillae

Transition zone

Columnar
epithelium with
goblet cells

Lamina
propria

Internal
hemorrhoidal plexus

Anal
sphincter
muscle

FIGURE 15.62

LM of a longitudinal section through the rectoanal junction at low magnification. The micrograph shows the transition of the epithelia from the simple columnar epithelium of the rectum to the stratified nonkeratinized squamous epithelium of the anal region. Intestinal glands are absent in the anal canal. The mucous membrane forms a series of longitudinal folds called the rectal columns of Morgagni. The lamina propria, muscularis mucosae, and submucosa are disorganized and difficult to differentiate. (100×)

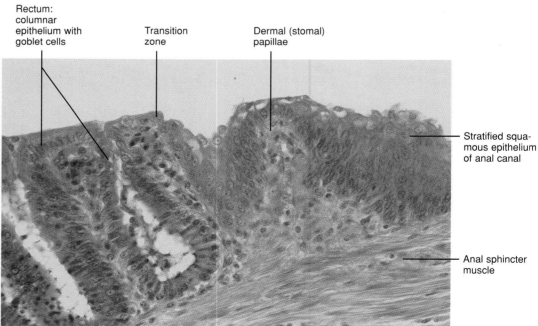

Rectum:
columnar
epithelium with
goblet cells

Transition
zone

Dermal (stomal)
papillae

Stratified squamous epithelium
of anal canal

Anal sphincter
muscle

FIGURE 15.63

LM of a section through the rectoanal junction, at a higher magnification. Shown is the region of transition from the columnar epithelium of the rectum to the stratified squamous epithelium of the anal canal. The mucosal goblet cells in the micrograph are associated with the rectum. (200×)

Liver

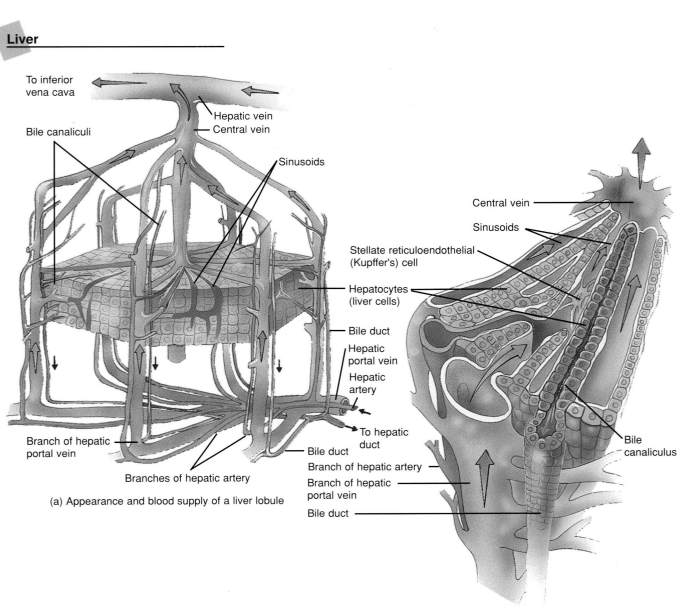

To inferior
vena cava

Bile canaliculi

Hepatic vein
Central vein

Sinusoids

Branch of hepatic
portal vein

Bile duct

Branches of hepatic artery

To hepatic
duct

Bile duct

(a) Appearance and blood supply of a liver lobule

Central vein

Sinusoids

Stellate reticuloendothelial
(Kupffer's) cell

Hepatocytes
(liver cells)

Bile duct

Hepatic
portal vein

Hepatic
artery

Branch of hepatic artery

Branch of hepatic
portal vein

Bile duct

Bile
canaliculus

(b) Portion of a liver lobule

FIGURE 15.64
(a) Three-dimensional diagram of a liver lobule, its associated blood
supply, and the sinusoids of the liver. (b) Diagram of a portal area,
hepatocytes, a few sinusoids, the bile canaliculus, and the central vein.

Kupffer's cells

FIGURE 15.65
LM of a cross section through the liver tissue at low magnification. Visible are the liver sinusoids leading to the central vein, endothelial cells lining the central vein, and trapped blood cells within the liver sinusoids. (100×)

Central vein

Hepatic sinusoids

Portal canal

FIGURE 15.66
LM of a cross section through a hepatic lobule. The micrograph shows the arrangement of hepatic parenchyma in which hepatocytes form flat anastomosing plates generally one cell thick. Also visible are blood cells and Kupffer's cells in the liver sinusoid. The large opening of the central vein with endothelial cells can also be identified. (200×)

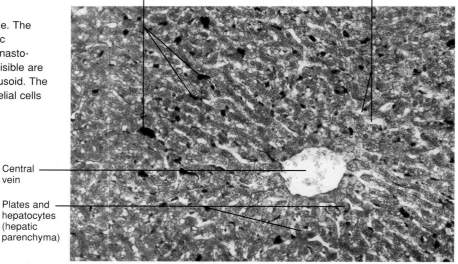

Kupffer's cells

Hepatic sinusoids

Central vein

Plates and hepatocytes (hepatic parenchyma)

Hepatic sinusoids

Kupffer's cells

Bi-nucleated hepatocytes

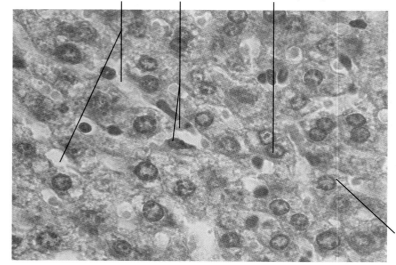

FIGURE 15.67
LM of a cross section of liver tissue, at a higher magnification, showing the cellular arrangement within the lobule. In this micrograph the hepatocytes can be identified by their large nuclei with peripherally arranged chromatin. As a general rule, almost 25% of the hepatocytes are binucleated. The other types of cells found in the lobules are the sinusoidal lining cells, which are identified by their flattened, condensed nuclei and scant clear cytoplasm, the phagocytosing Kupffer's cells in the sinusoids. These cells can be identified in the micrograph by the ingested blue dye particles in their cytoplasm. (1000×)

Mononucleated hepatocyte

Nuclei of
endothelial cells

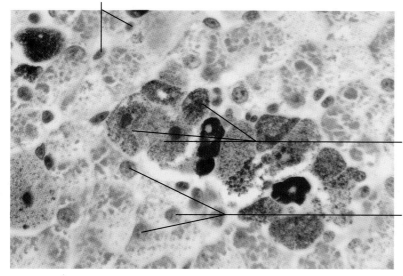

Glycogen
granules

Nuclei of
hepatocytes

FIGURE 15.68
LM of liver hepatocytes with stained, irregular
glycogen granule deposits of various sizes. Also
visible in the micrograph are sinusoidal cells with
condensed nuclei and hepatocytes that are devoid
of glycogen granules. (400×)

FIGURE 15.69
LM of liver tissue, showing minute bile canaliculi
between adjacent hepatocytes. The canaliculi are channels that collect bile from hepatocytes and drain it
toward the bile ducts located in the portal tracts. (400×)

Sinusoids

Bile
canalicula

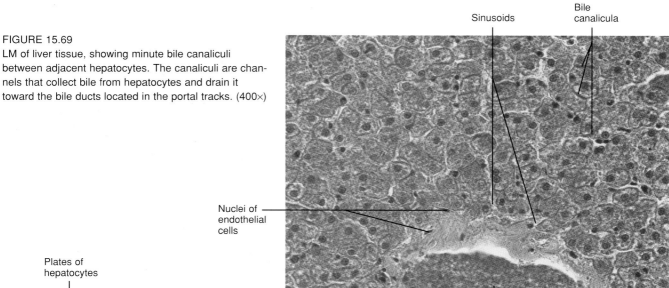

Nuclei of
endothelial
cells

Central vein with blood cells

Plates of
hepatocytes

Branch of
hepatic
artery

Branch
of portal
vein

Sinusoids

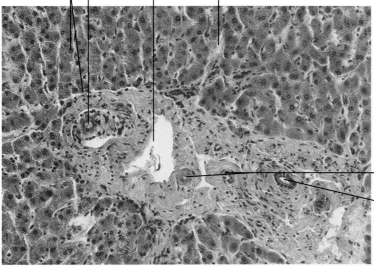

Branches of
hepatic artery

Bile duct

FIGURE 15.70
LM of a cross section through the portal tract between liver
lobules. The portal tract contains a branch of a hepatic
portal vein (generally with a large lumen), a small thick-
walled arteriole or an artery-like structure that branches off
the hepatic artery, and a bile-collecting duct that is internally
lined with simple cuboidal or low columnar cells. A fourth
vessel, the lymphatic, may also be visible if the body wall of
the vessel has not collapsed. Portal tracks are also called
portal areas. (200×)

Gallbladder

FIGURE 15.71
LM of a cross section through the gallbladder at a low magnification. The gallbladder is a sac-like structure with pronounced folds formed by the mucosa. It is lined by simple columnar epithelium. Smooth muscle cells in the body wall of the gallbladder can be stimulated to contract by cholecystokinin-pancreozymin (CCK), the hormone secreted by the duodenum. Loose connective tissue below the mucosa forms the submucosa. A muscle layer separates the submucosa from the adventitial connective tissue. (20×)

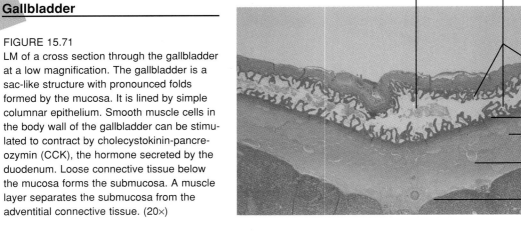

Lumen of the gallbladder

Pronounced folding of the mucosa

Lamina propria

Fibromuscular coat

Perimuscular connective tissue

Serosa

FIGURE 15.72
LM of a section of a gallbladder at a higher magnification. Tall simple columnar epithelium cells line the mucosa. Microvilli form the brush border of the epithelium. Also visible in the micrograph are the loose connective tissue of the submucosa and the underlying smooth muscle layer. (200×)

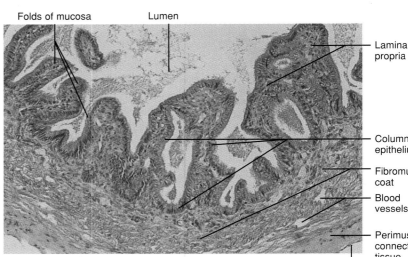

Folds of mucosa

Lumen

Lamina propria

Columnar epithelium

Fibromuscular coat

Blood vessels

Perimuscular connective tissue

Serosa

FIGURE 15.73
LM of a cross section through the common bile duct (ductus choledochus) before its juncture with the duodenal ampulla of Vater. The morphology of the common bile duct is similar to the body wall of the gallbladder, although on a smaller scale. Visible in the micrograph are the bile duct, the muscularis, the submucosa, and the body wall of the duodenum. (20×)

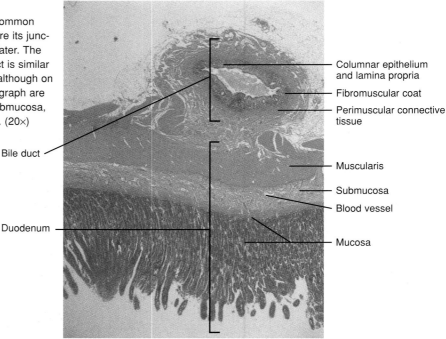

Columnar epithelium and lamina propria

Fibromuscular coat

Perimuscular connective tissue

Muscularis

Submucosa

Blood vessel

Bile duct

Mucosa

Duodenum

FIGURE 15.74
LM of a cross section through the bile duct (ductus choledochus), at a higher magnification. Tall simple columnar cells form the epithelium. The mucosa with many folds, the subepithelial tissue, the poorly organized lamina propria, and lymphoid tissue in the form of nodules can be observed at one end of the section. Well-organized smooth muscle cells form the outer wall of the duct near the duodenum. The lumen of this bile duct shows bile concentration. (100×)

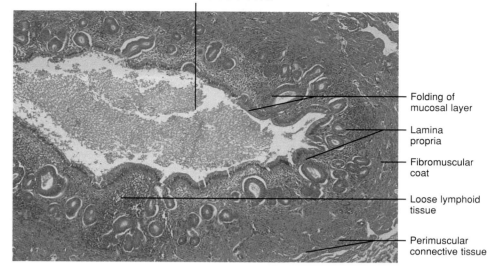

Lumen of the bile duct

Folding of mucosal layer

Lamina propria

Fibromuscular coat

Loose lymphoid tissue

Perimuscular connective tissue

Pancreas

FIGURE 15.75
LM of a section through the pancreas. The pancreas is both an endocrine and an exocrine gland. It is divided into many lobules that are separated by loose connective tissue septa. The exocrine part of the pancreas consists of densely packed secretory acini. The endocrine component of the pancreas forms clusters of cells various sizes called the islets of Langerhans. The islets are scattered throughout the tissue. Also visible in the micrograph are blood vessels and secretory ducts. (20×)

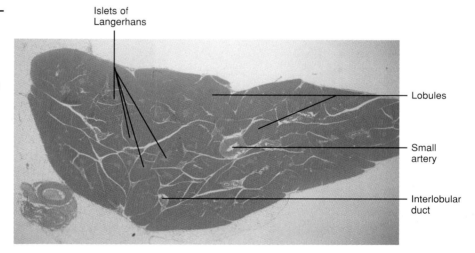

Islets of Langerhans

Lobules

Small artery

Interlobular duct

FIGURE 15.76
LM of a pancreas, illustrating the details of pancreatic serous acini or alveoli. As shown in the micrograph, the alveoli are tubular or pear-shaped, and there are generally five to eight cells in an acinus. The smaller cells in the acinus are centroacinar cells that are part of the duct system. These cells often originate in the central part of the acinus. (200×)

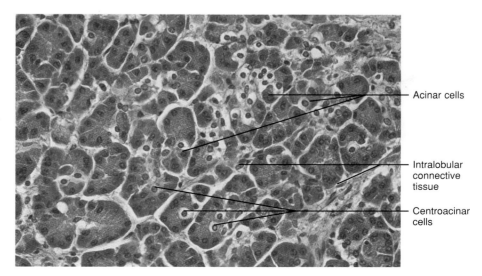

Acinar cells

Intralobular connective tissue

Centroacinar cells

FIGURE 15.77
LM of a cross section of pancreatic acini, at a higher magnification. As shown in the micrograph, acinar cells have spherical nuclei that lie toward the base of the cell. Large, spherical zymogen granules (acidophil secretion) in the cytoplasm of the cells have a limiting membrane. The acini are surrounded by loose connective tissue. (1000×)

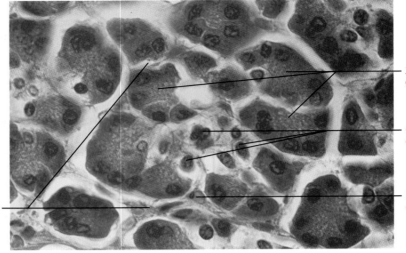

Zymogen granules

Centroacinar cells

Blood capillary

Intralobular connective tissue

FIGURE 15.78
LM of a cross section through the islet of Langerhans in the pancreas. Clusters of endocrine cells are supported by reticular connective tissue and blood vessels that are essentially fenestrated capillaries. A thin, fibrous capsule surrounds the islet. The surrounding acini cell granules stain much darker than the poorly stained islet granules and cytoplasm. (400×)

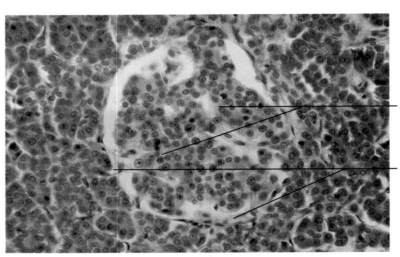

Islet of Langerhans cells

Capsule

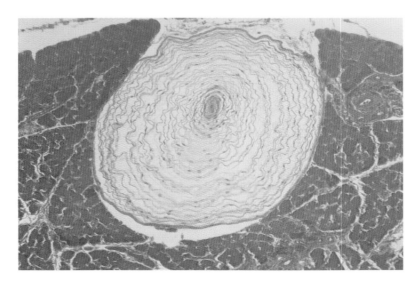

FIGURE 15.79
LM of a cross section through the pancreatic tissue in a cat. In the micrograph, a large encapsulated Pacinian corpuscle is surrounded by pancreatic acini. The function of such a capsule in the pancreas is not clear. (400×)

Respiratory System

The function of the **respiratory system** is to take in oxygen and eliminate carbon dioxide from the blood. Oxygen is utilized in cellular respiration, and carbon dioxide is a by-product of cellular metabolism. This gaseous exchange occurs on two levels: **external respiration** occurs in the lungs, and **internal respiration** occurs at the cellular level between cells and the blood. The exchange of gases is a diffusion process. Oxygen diffuses or moves from a higher concentration to a lower concentration, as in alveolar air. **Carbon dioxide** diffuses from the blood to the **alveolar** air, which has a lower concentration of carbon dioxide. Morphologically, the respiratory system can be divided into two parts: (1) the upper respiratory structures, including the **interconnected cavities—** the **nasal cavity,** the **paranasal sinuses,** the **nasopharynx,** and the **pharynx;** and (2) the lower respiratory tract, which includes the **larynx,** the **trachea,** the **lungs,** and the associated structures of **bronchi, bronchioles** and **alveolar ducts, alveoli,** and surrounding **blood vessels.**

The Upper Respiratory Tract—Nose and Nasal Cavity

A septum internally divides the nose into right and left **nasal cavities.** Within each cavity are three **turbinate bones** or **conchae.** The nasal cavities, including the conchae, are covered by respiratory epithelium supported by an underlying connective tissue, the lamina propria. The lamina propria is highly vascularized with thin-walled blood vessels and is called **cavernous** or **erectile** tissue. Also present in the connective tissue are mucous and serous glands. The epithelial lining of the nasal cavities is predominantly ciliated columnar cells. The **paranasal air sinuses** associated with the nasal cavities—**maxillary, frontal, ethmoidal,** and **sphenoidal**—act as resonance chambers for sound. The sinuses and the nose contain **receptors** for smell.

Nasopharynx The nasopharynx is located above the soft **palate** and behind the **posterior nares.** The epithelium in the area is pseudostratified ciliated columnar or stratified squamous, the latter being present at the junction of the posterior soft palate and the **pharynx**. The nasopharynx opens into the large **pharyngeal cavity**. The posterior and lateral walls of the pharynx are muscular to allow for the flexibility to dilate and constrict.

The Lower Respiratory Structures

Leading into the pharynx are the two **nasopharynges, one oropharynx, one laryngopharynx,** two openings for the **auditory (eustachian)** tubes, and one opening for the **esophagus**. Also present on the posterior wall of the pharynx is the **pharyngeal tonsil (adenoid).**

Larynx The larynx is a cartilaginous structure that connects the cavities of the **pharynx** and the **trachea**. The cartilages that comprise the body wall of the larynx are the **thyroid**, the **cricoid**, the **corniculates**, the **cuneiforms**, the **arytenoids**, and the **epiglottis**. The cartilages of the larynx are connected to the **hyoid bone** by three thin, flat membranes: the **thyrohyoid**, the **quadrate**, and the **cricoid**. The epithelium lining the larynx varies with location. It may contain nonkeratinized stratified squamous epithelium in the upper region and pseudostratified ciliated columnar epithelium with goblet cells in the lower region.

Trachea and Bronchi The trachea is a tube 10 to 12 cm long with about 20 spaced, horseshoe-shaped cartilaginous rings that keep the trachea from collapsing. At the lower end, the trachea bifurcates into the right and left main or **primary bronchi**. Externally, the trachea and the bronchi are surrounded by the **adventitia**. Beneath the adventitia lies the submucosa, a layer of loose **areolar connective** tissue with small serous glands. The innermost lining is the mucosa, which consists of pseudostratified ciliated epithelium mixed with goblet cells. The main bronchus enters the lung and divides into smaller bronchi, the **secondary** and **tertiary bronchi**. The tertiary bronchi supply air to ten segments in each lung.

Some form of cartilage is maintained in the body wall throughout the subdivision of the bronchi. The epithelial lining in the smaller bronchi changes from pseudostratified ciliated columnar epithelium (as seen in the larger bronchi) to ciliated columnar epithelium.

Bronchioles There is no abrupt transition in the body wall structure as tertiary bronchi differentiate into **bronchioles**. The bronchioles lack cartilage and glands in their body walls and are surrounded by only a thin **fibrous adventitia**. The lamina propria in the bronchioles has prominent smooth muscle bundles and elastic fibers. The epithelium is ciliated columnar mixed with goblet cells. As the bronchioles become smaller, the goblet cells disappear and the ciliated columnar cells take on the shape of low columnar or cuboidal epithelium.

Alveolar Ducts and Sacs The alveolar ducts and sacs (**alveoli**) are at the terminal ends of the respiratory tree. The alveolar ducts are narrow, thin-walled tubes lined by simple squamous epithelium. The tubular lining is so narrow that it is hard to resolve by light microscopy. The alveolar ducts open into a single alveolus or into clusters of alveolar sacs (clusters of alveoli). The alveoli are supported by elastic and reticular fibers, **capillary plexus**, simple squamous epithelial cells, **macrophages (dust cells), septal cells,** and a basement membrane.

Blood Supply and Blood Vessels of the Lungs The lungs have a double blood supply. In pulmonary circulation, the deoxygenated blood from the **right ventricle** of the heart is pumped into the lungs by means of the **pulmonary artery**. It is oxygenated and returned to the **left auricle** of the heart by the pulmonary vein. The pulmonary arteries are large, thin-walled elastic arteries. The expansion and recoil of the elastic fibers in the body walls of the arteries maintains a continuous flow of blood through the lungs.

The second blood flow system through the lungs is maintained by the **bronchial arterial** system. Small branches of blood vessels originating from the **aorta** enter the lungs and supply oxygenated blood to the tissues of the lungs and the **pleura**. A small proportion of deoxygenated blood is collected by the bronchial system and returned to the right atrium by means of the **azygos venous system.**

Respiratory System

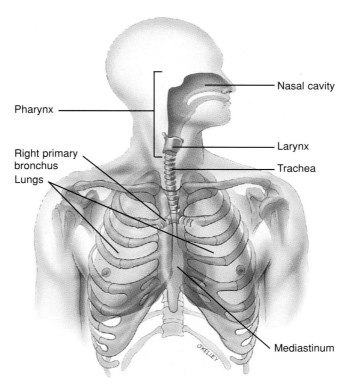

Anterior view

FIGURE 16.1
Diagram of organs of respiration and their relationship to the surrounding structures. The figure shows the location of the larynx, the trachea, and the lungs in the thorax.

Upper Respiratory System

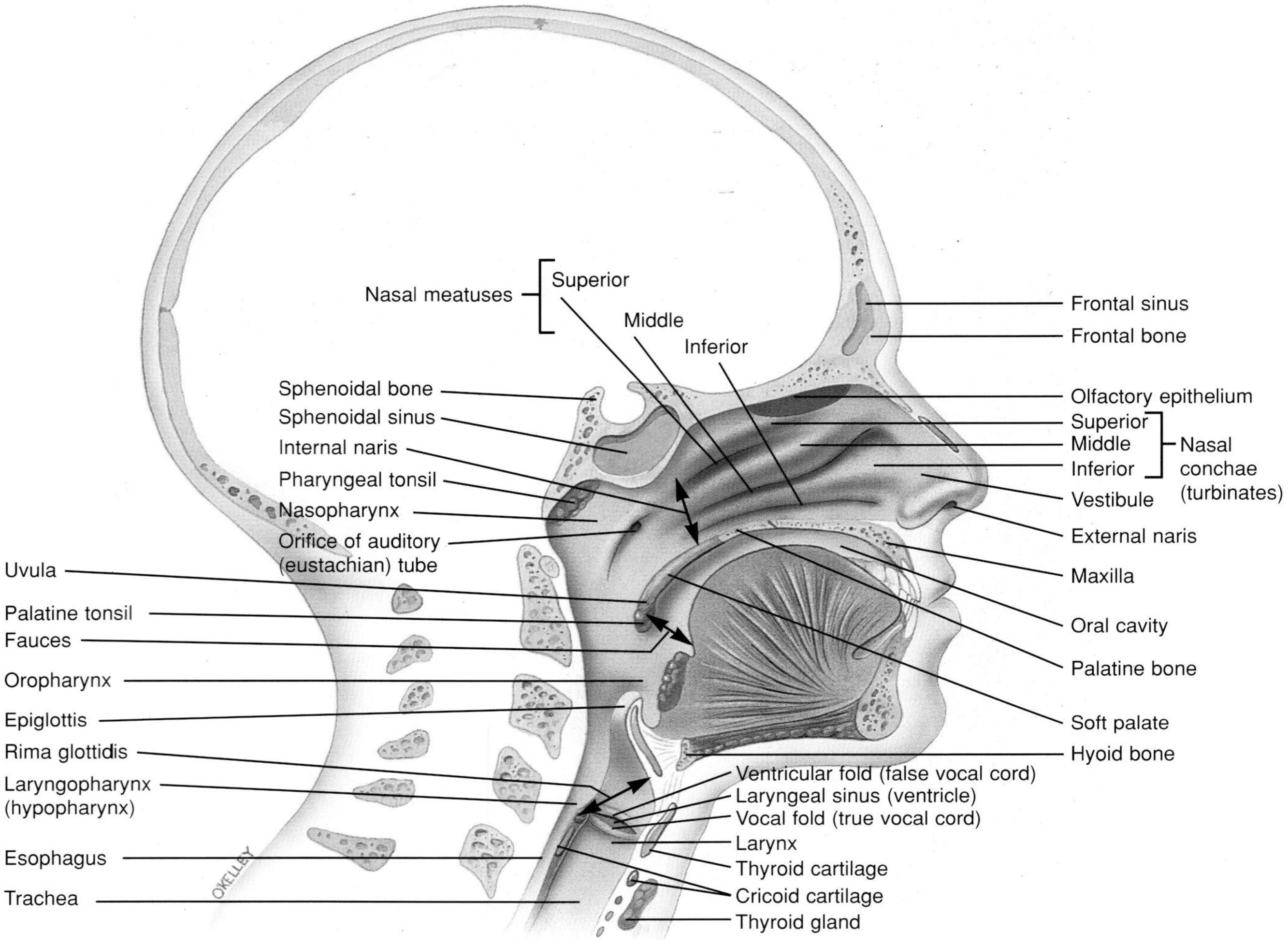

FIGURE 16.2
Diagram of a sagittal section through the left side of the head and neck region. The diagram identifies some of the structures associated with the upper left side of the respiratory system.

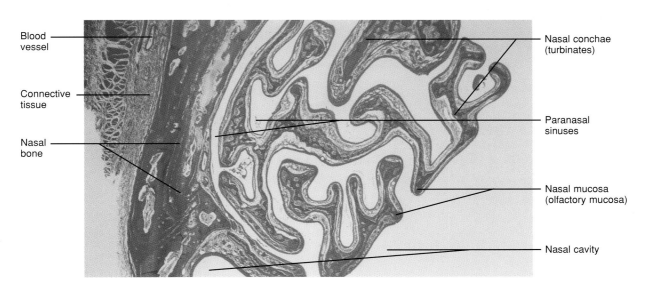

Blood vessel

Connective tissue

Nasal bone

Nasal conchae (turbinates)

Paranasal sinuses

Nasal mucosa (olfactory mucosa)

Nasal cavity

FIGURE 16.3
Light micrograph (LM) of a frontal section through the upper part of a nasal cavity. Visible in the micrograph are the nasal conchae (turbinates) lined by nasal mucosa. The conchae are projecting into the nasal cavity. Bone and connective tissue can also be identified in the micrograph. (20×)

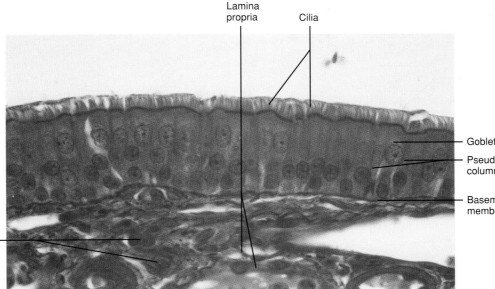

Lamina propria

Cilia

Goblet cell

Pseudostratified columnar epithelium

Basement membrane

Serous and mucous glands

FIGURE 16.4
LM of a transverse section through the nasal mucosa. Pseudostratified ciliated columnar epithelium with goblet cells, basement membrane, and underlying lamina propria connective tissue can be identified in the micrograph. (1000×)

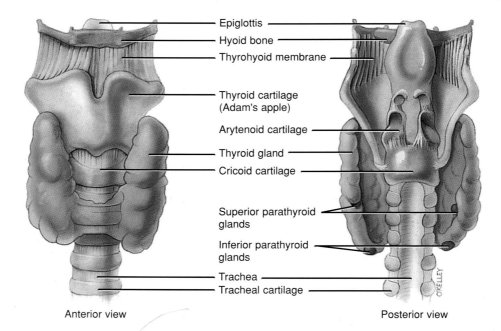

Epiglottis
Hyoid bone
Thyrohyoid membrane

Thyroid cartilage
(Adam's apple)

Arytenoid cartilage

Thyroid gland
Cricoid cartilage

Superior parathyroid
glands

Inferior parathyroid
glands

Trachea
Tracheal cartilage

Anterior view   Posterior view

FIGURE 16.5
Diagram of the larynx and its cartilages, the trachea, and the thyroid gland on the larynx and the trachea. Relative locations are illustrated.

FIGURE 16.6
LM of a longitudinal section through the epiglottis. The micrograph shows the lingual and pharyngeal surfaces, which are covered with nonkeratinized stratified squamous epithelium. Underlying the epithelium is the lamina propria with serous and mucous glands. Also visible is the elastic cartilage that forms the central core of the epiglottis, and the lymphoid nodules at the base of the epiglottis. (20×)

Mucosa of the
anterior surface

Stratified squamous
epithelium

Mucosa posterior
laryngeal surface

Elastic cartilage

Perichondrium

Blood vessel

Lymphoid
nodule

Mixed glands in
lamina propria

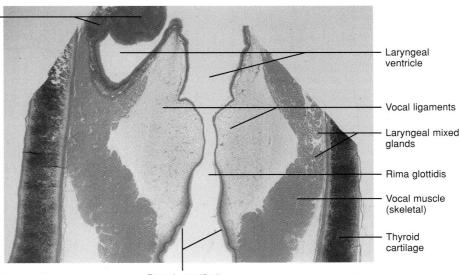

Lymphoid nodules

Laryngeal ventricle

Vocal ligaments

Laryngeal mixed glands

Rima glottidis

Vocal muscle (skeletal)

Thyroid cartilage

Pseudostratified ciliated columnar epithelium

FIGURE 16.7
LM of a frontal section through the larynx in the upper region. The epithelial lining of the larynx is nonkeratinized stratified squamous epithelium. Below the epithelium lies specialized fibro-elastic connective tissue that forms the vocal ligament. Vocalis muscle (skeletal muscle), a ventricle, a lymphoid nodule, the lamina propria, perichondrium, and thyroid cartilage are also visible. (20×)

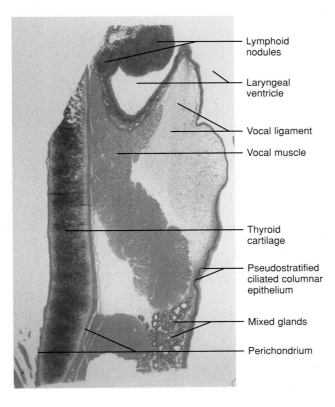

Lymphoid nodules

Laryngeal ventricle

Vocal ligament

Vocal muscle

Thyroid cartilage

Pseudostratified ciliated columnar epithelium

Mixed glands

Perichondrium

FIGURE 16.8
LM of a frontal section of the larynx, showing the locations of mixed glands. Identifiable in the micrograph are a lymphoid nodule, nasal epithelium, underlying lamina propria of fibroelastic connective tissue that forms the vocal ligaments, vocal muscle, vestibule, perichondrium, thyroid cartilage, and the mixed gland found in the lower part of the larynx. (20×)

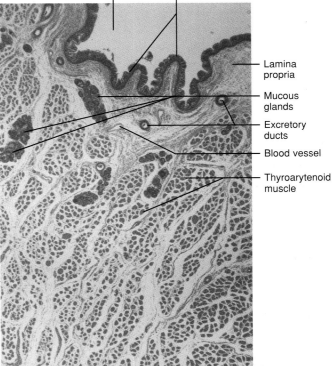

Laryngeal ventricle

Stratified squamous epithelium

Lamina propria

Mucous glands

Excretory ducts

Blood vessel

Thyroarytenoid muscle

FIGURE 16.9
LM of a frontal section through a small area of the larynx, at a higher magnification. Mucosal nonkeratinized stratified squamous epithelium, the lamina propria with fibroelastic connective tissue, mucous glands and excretory ducts, blood vessels, thyroarytenoid muscle, and the laryngeal ventricle are discernible. (40×)

FIGURE 16.10
LM of the epithelial lining of the larynx below the vocal
folds. The lining in this region changes from stratified
squamous to pseudostratified ciliated columnar epithelium mixed with mucus-secreting goblet cells. Beneath
the epithelium lies the lamina propria with connective
tissue fibers and adipose cells. (1000×)

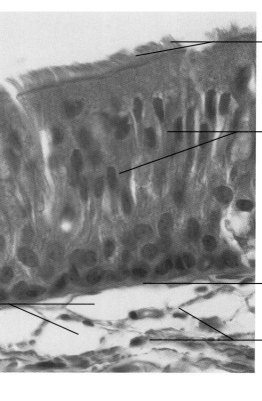

Cilia

Pseudostratified
ciliated columnar
epithelium

Basement
membrane

Adipose
cells

Lamina
propria

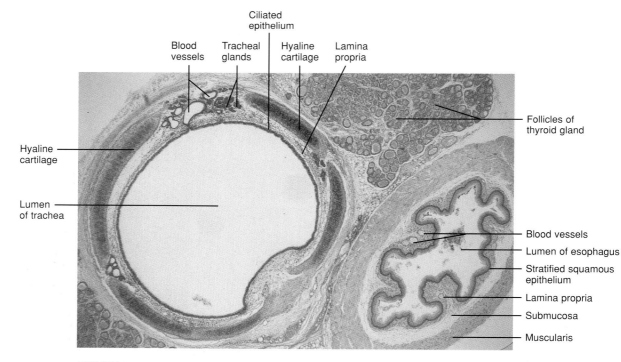

Ciliated
epithelium

Blood
vessels

Tracheal
glands

Hyaline
cartilage

Lamina
propria

Hyaline
cartilage

Lumen
of trachea

Follicles of
thyroid gland

Blood vessels

Lumen of esophagus

Stratified squamous
epithelium

Lamina propria

Submucosa

Muscularis

FIGURE 16.11
LM of a cross section through the trachea and the esophagus. The trachea located anteriorly has a large lumen with C-shaped hyaline cartilage ring in the body wall. The esophagus, when empty, has a smaller lumen. The mucosa of the trachea is lined by nonkeratinized stratified squamous epithelium. Beneath the epithelium is the lamina propria,
submucosa, muscularis, and adventitia. (40×)

FIGURE 16.12

LM of a cross section through the trachea. The micrograph shows the lumen at the top. The lining of the trachea is pseudostratified ciliated columnar epithelium mixed with mucus-secreting goblet cells. Below the epithelium is the basement membrane and lamina propria with elastic connective tissue. The submucosa contains serous mucus-secreting glands. Beneath the submucosa is the C-shaped cartilaginous ring. (100×)

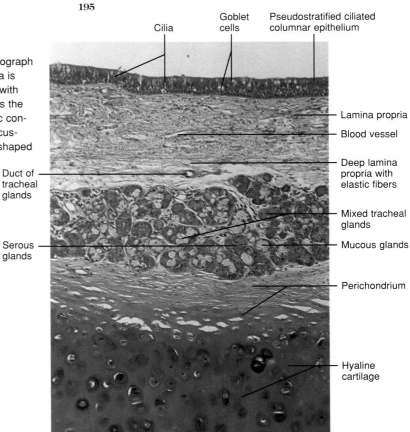

Cilia

Goblet cells

Pseudostratified ciliated columnar epithelium

Lamina propria

Blood vessel

Deep lamina propria with elastic fibers

Duct of tracheal glands

Mixed tracheal glands

Mucous glands

Serous glands

Perichondrium

Hyaline cartilage

FIGURE 16.13

LM of a cross section through the mucosal lining of the trachea. The trachea and the bronchi have similar mucosal linings. The linings will be discussed collectively as tracheobronchial epithelium. The tracheobronchial epithelium of the mucosal layer is lined with pseudostratified ciliated columnar epithelium with mucus-secreting goblet cells. Two other types of cells, mixed with the ciliated and goblet cells, can be identified at the electron microscopic level. These are brush cells with microvilli and basal cells. The basal cells are believed to be stem cells that later differentiate into ciliated cells. The function of the brush cells is unknown. (1000×)

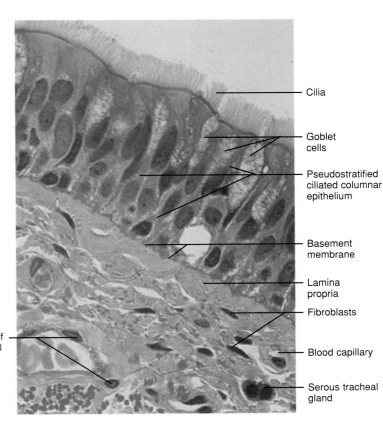

Cilia

Goblet cells

Pseudostratified ciliated columnar epithelium

Basement membrane

Lamina propria

Fibroblasts

Blood capillary

Ducts of tracheal glands

Serous tracheal gland

FIGURE 16.14

LM of a cross section through an intrapulmonary bronchus, which is a relatively large passage for air. The body wall is thick because of the presence of irregular hyaline cartilaginous plates. The plates are surrounded by fibroelastic connective tissue. Interspersed in the connective tissue are mucoserous and mucous glands, and interlacing bundles of smooth muscle fibers. The mucosa is lined by ciliated epithelium, and the lamina propria contains elastic fibers. The mucosa shows various degrees of longitudinal folds resulting from contraction of the smooth muscle. (100×)

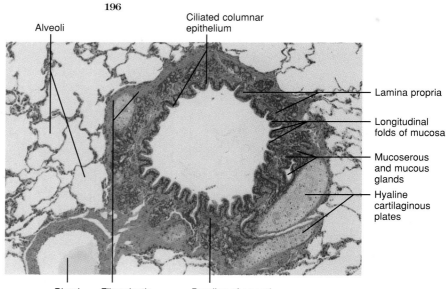

Alveoli

Ciliated columnar epithelium

Lamina propria

Longitudinal folds of mucosa

Mucoserous and mucous glands

Hyaline cartilaginous plates

Blood vessel

Fibroelastic connective tissue

Bundles of smooth muscle fibers

FIGURE 16.15

LM of a cross section through the lung tissue, illustrating a bronchiole, a bronchus, and a branch of the pulmonary artery. The bronchiole is a conducting tube with respiratory tissue surrounding the body wall. In the body wall of the bronchiole there is complete absence of cartilage. The lamina propria is of elastic fibers that surround smooth muscle. The epithelium is ciliated columnar with some goblet cells. In smaller bronchioles there is an absence of goblet cells and the ciliated cells are low columnar or occasionally cuboidal in shape. With the ciliated cells are a few chiolar or Clara cells that project into the lumen of the bronchiole. (200×)

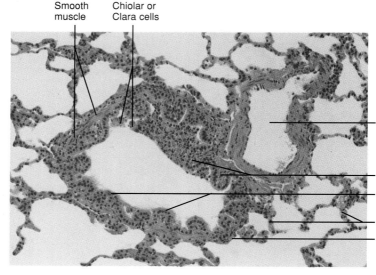

Smooth muscle

Chiolar or Clara cells

Branch of pulmonary artery

Lamina propria

Ciliated columnar epithelium

Alveolar wall

Adventitia

Lamina propria with smooth muscle and fibroelastic connective tissue

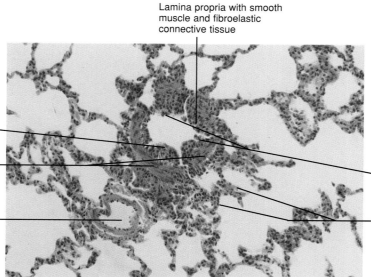

FIGURE 16.16

LM of a small bronchiole in cross section. The epithelial lining displays nonciliated cuboidal cells with an absence of goblet cells. Protruding into the lumen are chiolar or Clara cells (bronchiolar cells). Bundles of smooth muscle cells are prominent in the lamina propria. (10×)

Cuboidal epithelial cells

Smooth muscle

Branch of pulmonary artery

Chiolar or Clara cells

Alveolar wall

Lower Respiratory System

FIGURE 16.17
Diagram of a portion of a lung lobule.
Shown are the terminal structures of the
respiratory tree, the terminal bronchiole,
respiratory bronchiole, alveolar ducts, and
alveolar sacs.

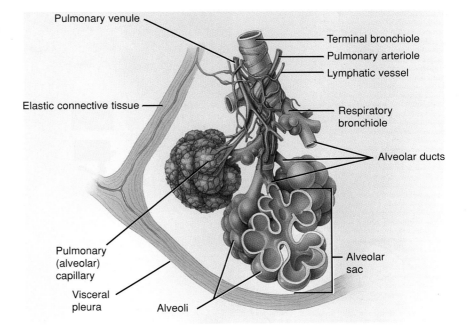

Pulmonary venule
Terminal bronchiole
Pulmonary arteriole
Lymphatic vessel
Elastic connective tissue
Respiratory bronchiole
Alveolar ducts
Pulmonary (alveolar) capillary
Alveolar sac
Visceral pleura
Alveoli

FIGURE 16.18
LM of a cross section through lung tissue
with a terminal bronchiole. The epithelium
shows patches of low ciliated columnar
cells interspersed with cuboidal cells.
Smooth muscle fibers mixed with elastic
fibers can be seen in the lamina propria.
Leading from the terminal bronchiole are
conduits (passages) to the respiratory
bronchiole. (100×)

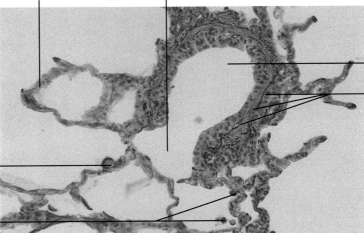

Alveolar wall
Respiratory bronchiole
Terminal bronchiole
Smooth muscle and elastic fibers of lamina propria
Blood capillary
Dust cells (macrophages)

FIGURE 16.19
LM of respiratory bronchiole in cross section. The epithelial lining is ciliated cuboidal
epithelium, and is continous with simple
squamous epithelium. Interlacing smooth
muscle bundles and fibroconnective tissue
can be seen at places in the body wall of
the bronchiole. (200×)

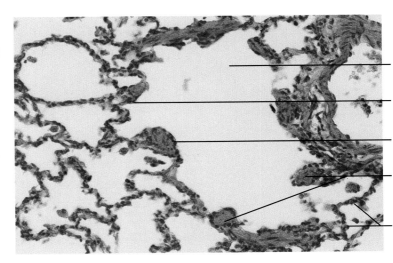

Respiratory bronchiole
Simple squamous epithelium
Simple cuboidal
Smooth muscle and fibroconnective tissue
Alveolar wall

Respiratory bronchiole

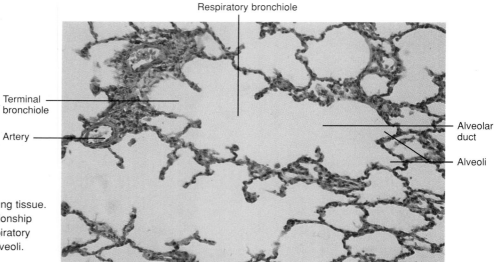

Terminal bronchiole

Artery

Alveolar duct

Alveoli

FIGURE 16.20
LM of a cross section through lung tissue. The micrograph shows the relationship among terminal bronchiole, respiratory bronchiole, alveolar duct, and alveoli. (200×)

Blood capillaries

Branch of pulmonary artery

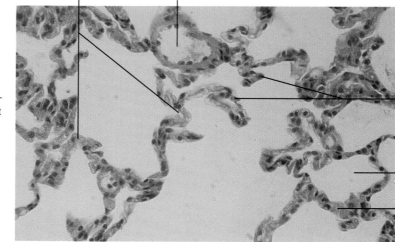

FIGURE 16.21
LM of lung tissue in cross section. The micrograph shows several alveoli and their supporting body walls. The body walls consist of simple squamous epithelium, septal cells, type I and type II pneumocytes, endothelial cells of the capillaries, and alveolar macrophages or dust cells. (100×)

Epithelial cells (simple squamous)

Alveolus

Elastic fibers

Endothelial cell

Squamous epithelial cell (pneumocyte I)

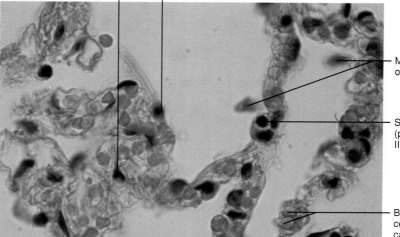

FIGURE 16.22
LM of lung tissue in cross section, at a higher magnification. The micrograph shows several alveoli and their supporting body walls. Simple squamous epithelial cells (pneumocyte I) and surfactant cells (pnemocyte II or septal cells) line the body walls. The surfactant cells secrete a fluid that inhibits the surface tension on the outer walls of the alveoli. Endothelial cells of the blood capillaries and phagocytic macrophages or dust cells can also be identified in the micrograph. (1000×)

Macrophages or dust cells

Surfactant cell (pneumocyte II)

Blood cells in capillary

Urinary System

Cellular metabolic waste products, especially **nitrogenous compounds**, excess **electrolytes, water,** and other **toxic** substances, are eliminated by the **urinary system.** The process of excretion and urine formation involves the filtration of blood plasma by the kidneys. The filtration within the kidneys is brought about by filtration units called **nephrons.** The kidneys are able to function because they have an extremely efficient blood supply and a steady blood pressure. The blood pressure is maintained by **renin,** a hormone that is produced by the **juxtaglomerular cells (JG cells)** of the kidney. Kidneys are also essential to the secretion of **erythropoietin,** a hormone that stimulates bone marrow erythrocyte production.

The human kidneys are about 10 to 12 cm in length and 3.5 to 5 cm in width. They are enclosed in a fibrous capsule and are located in the upper posterior part of the abdomen, one on each side of the upper lumbar vertebrae. The **hilus (hilum),** a shallow depression on the medial aspect of the kidney, forms the entry point for the blood vessels. Exiting from the hilus is an excretory duct, the

ureter, which transports waste products and excess water from the kidneys to the urinary bladder. The proximal end of the ureter, the **pelvis,** is broad and fills the hilus of the kidney. Within the kidney, the pelvis divides into cup-like structures, two **major calyces** and 8 to 12 **minor calyces.** Each calyx minor surrounds a **renal papilla.** Opening into each papilla are 10 to 25 **collecting ducts.** The **medullary pyramids,** composed of parallel blood vessels and tubules, are also in the kidneys.

The **medulla** of the kidney can be grossly divided into outer and inner zones, reflective of the morphological differences in the walls of the tubules in the cortex and the medulla region. In the medulla, the pyramids are separated by cortical material that extends between the pyramids and renal columns (of **Bertin**).

Kidney Lobes and Lobules

A lobe consists of a pyramid and its overlying cortex. Because each kidney has several pyramids, it is **multilobular** or **multipyramidal.** A kidney lobule is a smaller functional unit that includes a **medullary ray** (cortical), nephrons that drain into the cortical ray, and the extension of the ray into the medullary pyramid.

Uriniferous Tubules

Each kidney contains many **uriniferous tubules.** Functionally, a uriniferous tubule is a continuous tubule that can be divided into a nephron (3 to 4 cm long) and a collecting tubule (2 cm long). The nephron is responsible for the filtration of blood and the formation of a filtrate. The collecting tubule functions as an excretory duct that collects filtrate from many nephrons and transports it in the form of urine to the **renal pelvis.**

Blood Supply to the Kidneys

The **renal artery,** a branch of the abdominal aorta, supplies blood to the kidneys. The renal artery enters the kidney at the hilus and divides into **segmental arteries** that later branch into smaller **interlobar arteries.** The interlobar arteries divide into **arcuate arteries** in the **corticomedullary border** area. The arcuate arteries ascend toward the cortex and branch off into **interlobular arteries.** Close to the Bowman's capsule of the nephron, the interlobular arteries give rise to **afferent arterioles** of the **glomeruli.** The **efferent arterioles** receive blood from the glomeruli, which in turn pass the blood to the **vasa recta** and **peritubular** capillaries. Blood exits the kidney through the **renal vein** after passing through **interlobular, arcuate, interlobar,** and **segmental veins.**

Nephrons

There are over a million nephrons in each kidney. A nephron is a long, highly convoluted tubular structure that starts blindly as a **Bowman's capsule** and terminates by joining an **excretory duct.** The filtration of blood takes place in the Bowman's capsule. The filtrate passes into the tubule, where selective absorption begins and continues throughout the tubule. Finally, the filtrate is altered to form **urine,** which is excreted by the collecting tubules into the **renal pelvis.**

Collecting Tubules or Excretory Ducts

The collecting tubules or excretory ducts are not considered part of the nephron. The distal convoluted tubule of the nephron is connected to the collecting tubule. Several nephrons connect to a single collecting tubule, which passes into a **medullary ray** and extends downward toward the **medulla.** Several collecting tubules join in the medulla to form a large **papillary duct** (of **Bellini**) that opens into the apex of a **papilla.** The papilla has a sieve-like appearance because of the numerous papillary duct openings. This region of the medulla is called the **area cribrosa.**

The collecting tubules collect urine from the nephrons and excrete it into a **calyx minor.** From there, the urine passes into a **calyx major,** and then to the **renal pelvis.** The **antidiuretic hormone (ADH)** controls the absorption of water in the collecting tubules. The urine is conveyed from the renal pelvis to the **ureters.** By the muscular contraction of the ureters, the urine is propelled into the **urinary bladder,** where it is stored and later excreted by the process of **micturition.**

Ureters

The ureters are tubular structures, approximately 25 to 30 cm in length, that connect the renal pelvis to the urinary bladder. The ureters have a well-established **mucosa** with **transitional epithelium** supported by a **basement membrane** and **lamina propria.** Below the mucosa lies the thick **muscularis,** which consists of smooth muscle cells separated by connective tissue. This smooth muscle tissue forms an inner longitudinal layer and an outer circular layer (opposite to what is seen in the gastrointestinal tract). A third oblique muscle layer is present at the distal end of the ureter, where the circular muscle layer is absent. The outermost layer of fibroelastic adventitia covers the muscularis; a loosely arranged peritoneum surrounds the adventitia.

Urinary Bladder

The body wall of the bladder is similar to the ureter in cross section. However, the **transitional epithelium** of the mucosa is thicker (6 to 8 layers when relaxed, and 2 to 3 layers when distended) in the bladder. The lamina propria is poorly organized in the bladder. Mucus-secreting cells are prevalent in the lamina propria, becoming more dense toward the urethral end. Below the lamina propria lies the **muscularis,** with smooth muscle fibers arranged in an inner **longitudinal** layer, a middle **circular** layer, and an outer **longitudinal** layer. A fibroelastic adventitia covers the muscularis. Peritoneum covers only the upper surface of the bladder.

Urethra

The terminal tubular structure of the urinary system is the urethra, with marked differences between the sexes.

In the male, the urethra is approximately 15 to 20 cm long. Functionally, the urethra is divided into **pars prostatica, pars membranacea,** and **pars cavernosa** or **pars spongiosa.** The pars prostatica is the first section, beginning at the orifice at the bladder and traversing through the prostate gland to open into the membranous urethra (pars membranacea). Two **ejaculatory ducts** and the ducts

from the **prostate gland** open in this region. The short pars membranacea traverses from the apex of the prostate through the skeletal muscle of the pelvis and the **perineal membrane** and terminates in the bulb of the **corpus cavernosa. The pars cavernosum** or **pars spongiosa** is the distal end of the urethra; it passes through the corpus spongiosum and opens into the **glans penis.** The epithelial lining of the **prostatic** region is **transitional** but changes to **pseudostratified** or **stratified columnar** in the remainder of the urethra. Patches of stratified squamous epithelium may also be present. Branching tubular glands (of **Littre**) are found in the deep depression of the urethral mucosa. These glands are more concentrated on the dorsal surface of the penile urethra.

In the female, the urethra is much shorter, approximately 4 cm in length. The muscularis is similar to that of the ureter. A striated muscle sphincter is located at the orifice. Stratified squamous epithelium with patches of pseudostratified or stratified columnar epithelium lines the urethra. The outpocketing of glands from the epithelium is similar to the **glands of Littre** found in the male. Fibroelastic connective tissue characterizes the lamina propria, in which numerous venous sinuses are present. These sinuses are similar to the cavernous tissue of the male urethra.

Kidney Lobes and Lobules

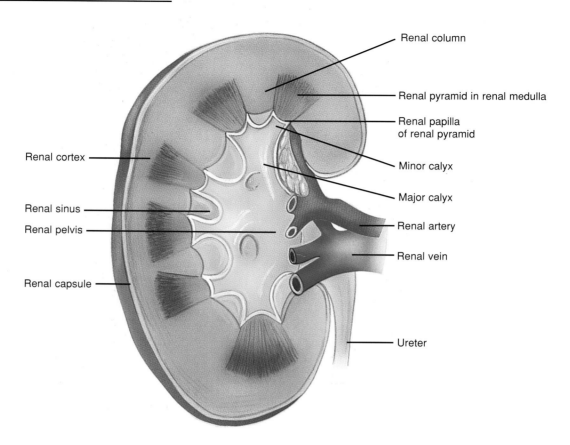

Renal column

Renal pyramid in renal medulla

Renal papilla
of renal pyramid

Minor calyx

Major calyx

Renal artery

Renal vein

Ureter

Renal cortex

Renal sinus

Renal pelvis

Renal capsule

FIGURE 17.1

Diagram of a coronal section of the kidney, illustrating the locations of
the cortex, medulla, ureter, renal artery, renal vein, and other salient
features associated with the kidney.

FIGURE 17.2
Light micrograph (LM) of a coronal section of the kidney, depicting the renal cortex, renal pyramids, and the pelvis. The hilus and the lumen of the renal artery are also identified in the diagram. (1×)

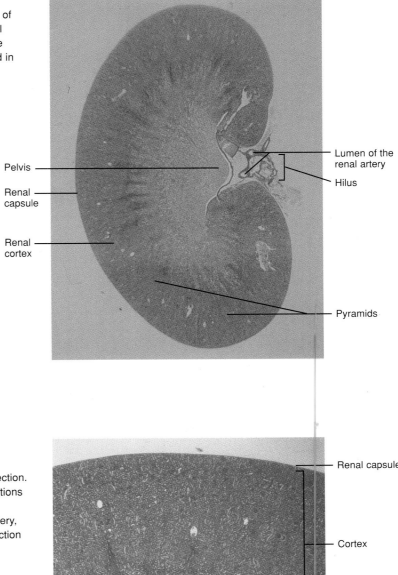

Pelvis

Renal capsule

Renal cortex

Lumen of the renal artery

Hilus

Pyramids

FIGURE 17.3
LM of a portion of the kidney in coronal section. Visible in the micrograph are the demarcations of the cortex, the medulla, and the renal papilla. In addition, a branch of a renal artery, the pelvis, and part of the renal vein in section are visible. (20×)

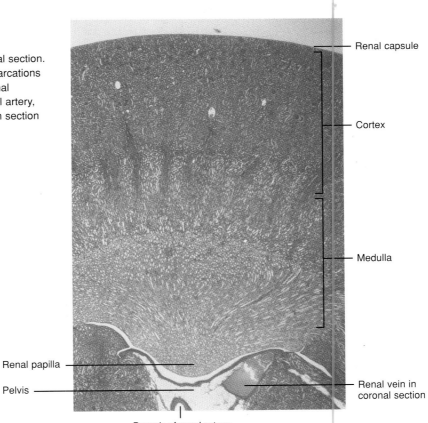

Renal capsule

Cortex

Medulla

Renal papilla

Pelvis

Renal vein in coronal section

Branch of renal artery

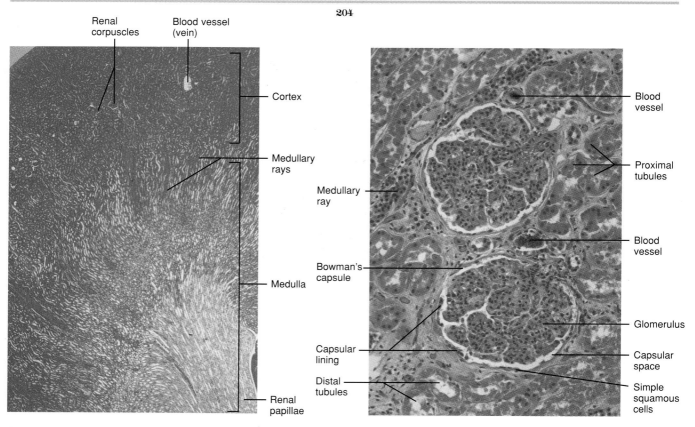

FIGURE 17.4

LM of a small portion of the kidney in coronal section. Visible are the cortex, the medulla, the medullary rays, and the renal papilla, as well as blood vessels and renal corpuscles. (20×)

FIGURE 17.5

LM of a kidney section with renal corpuscles (Bowman's capsules) and the surrounding tissue. The micrograph shows a medullary ray, proximal and distal tubules, blood vessels, glomeruli in the corpuscles, the simple squamous cell lining of the Bowman's capsule, the lining of the corpuscle, and capsular space. (200×)

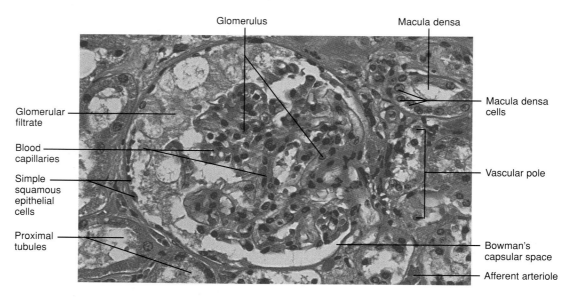

FIGURE 17.6

LM of a renal corpuscle (Bowman's capsule), illustrating the renal glomerulus, the capsular space, the simple squamous epithelium of the capsule, the blood capillaries surrounding the glomerulus, the filtrate in capsular space, the macula densa cells of the distal tubules, and the surrounding proximal tubules in cross section. (400×)

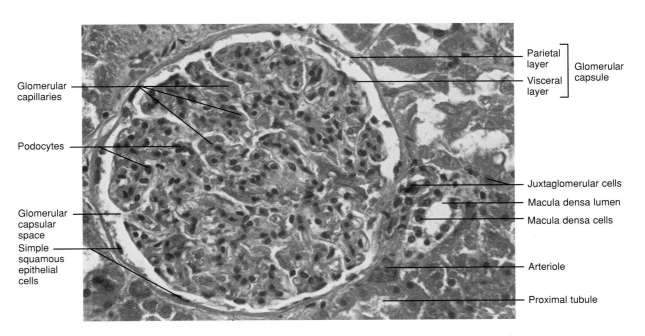

Glomerular capillaries

Podocytes

Glomerular capsular space

Simple squamous epithelial cells

Parietal layer
Visceral layer
Glomerular capsule

Juxtaglomerular cells

Macula densa lumen

Macula densa cells

Arteriole

Proximal tubule

FIGURE 17.7
LM of a section of renal corpuscle. The micrograph shows proximal tubules in section, the macula densa, capsular space, glomerular capillaries, afferent and efferent arterioles, visceral and parietal epithelial layers, podocytes, of the corpuscle. (400×)

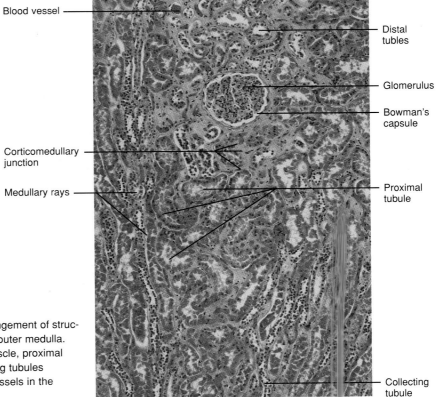

Blood vessel

Corticomedullary junction

Medullary rays

Distal tubles

Glomerulus

Bowman's capsule

Proximal tubule

Collecting tubule

FIGURE 17.8
LM of the kidney, illustrating the arrangement of structures associated with the cortex and outer medulla. The micrograph shows a renal corpuscle, proximal and distal tubules in section, collecting tubules forming medullary rays, and blood vessels in the area. (100×)

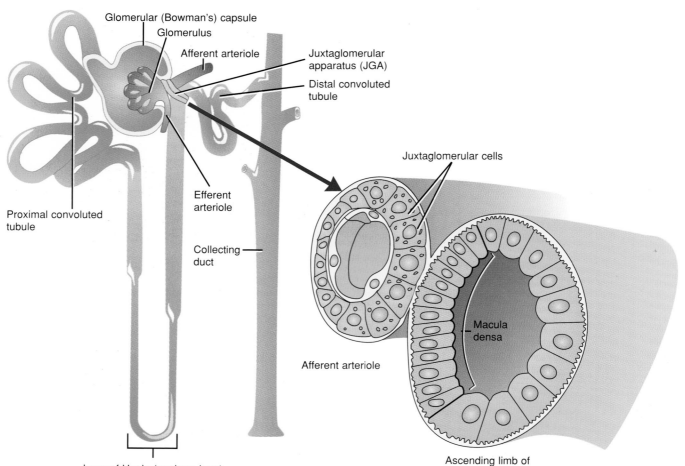

Glomerular (Bowman's) capsule

Glomerulus

Afferent arteriole

Juxtaglomerular apparatus (JGA)

Distal convoluted tubule

Proximal convoluted tubule

Efferent arteriole

Collecting duct

Loop of Henle (nephron loop)

Juxtaglomerular cells

Afferent arteriole

Macula densa

Ascending limb of the loop of Henle

FIGURE 17.9

(a) Diagram of a nephron and the collecting duct, showing the relationship between the afferent and efferent arterioles and the renal corpuscle. Also illustrated is the extension of the tubule into the medulla region to form the loop of Henle. (b) Cross-sectional view of the cells in the macula densa of the distal tubule and its closeness to the afferent arteriole.

FIGURE 17.10
LM of the cortical tissue of the kidney, in cross section. The micrograph shows the proximity of the renal corpuscle to the macula densa. Proximal tubules can be identified by their simple cuboidal cells and their tall microvilli that form brush borders in the lumen. Also shown are the distal convoluted tubules lined by cuboidal cells. However, these cells lack microvilli or brush borders. A large lumen of a collecting tubule can be seen at the bottom of the micrograph. (400×)

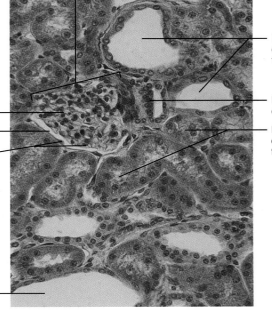

Renal corpuscle

Distal convoluted tubules

Macula densa cells

Proximal convoluted tubules

Glomerulus

Capsular space

Bowman's capsule

Collecting tubule

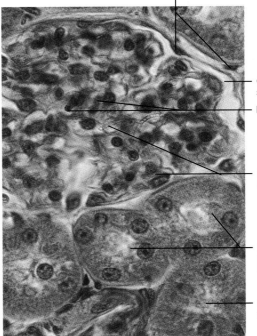

Bowman's capsule (glomerular capsule)

Capsular space

Podocytes

Glomerular capillaries

Microvilli of proximal tubules

Proximal tubule

FIGURE 17.11
LM of a renal corpuscle and proximal tubules at a higher magnification. The micrograph shows the simple cuboidal cells of the proximal tubules with microvilli extending into the lumen, which gives the appearance of a brush border. Parietal and visceral cells of the renal corpuscle (Bowman's capsule), blood capillaries, and podocytes surrounding the glomus can also be seen. (1000×)

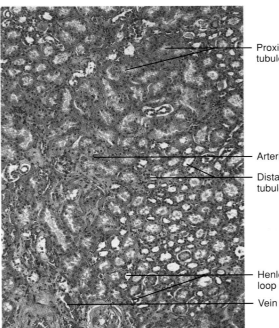

Proximal tubules

Arteriole

Distal tubule

Henle's loop

Vein

FIGURE 17.12
LM of the renal medulla, displaying the proximal and distal tubules, and Henle's loop in cross section. Also visible are small arteries and small veins. The Henle's loop in section can be distinguished from veins and arterioles by the rounded shape and lack of smooth muscle cells. The lumen is lined with simple squamous cells. (100×)

Henle's loop in
cross section

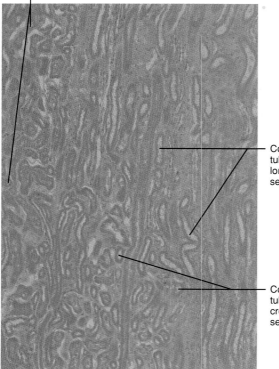

Collecting
tubules in
longitudinal
section

Collecting
tubules in
cross
section

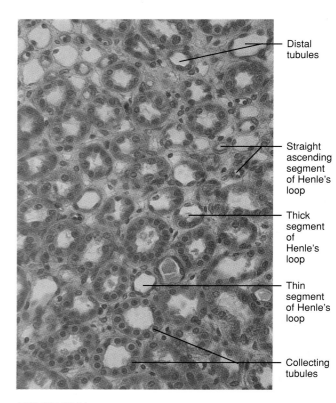

Distal
tubules

Straight
ascending
segment
of Henle's
loop

Thick
segment
of
Henle's
loop

Thin
segment
of Henle's
loop

Collecting
tubules

FIGURE 17.13
LM of the renal medulla, displaying the distal collecting tubules in longitudinal and cross section.
Henle's loop can be seen in cross section in the
upper part of the micrograph. (200×)

FIGURE 17.14
LM of the renal medulla in cross section. Shown in
the micrograph are distal tubules, thick and thin sections of Henle's loop, and collecting tubules. (200×)

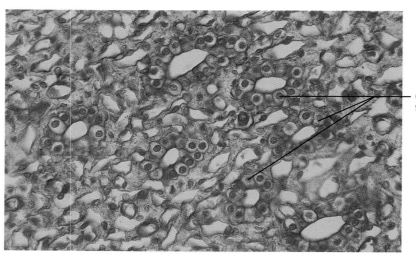

Collecting
tubules

FIGURE 17.15
LM of collecting tubules in cross section. As the collecting tubule approaches the
collecting duct, the cuboidal cells of the tubule become taller. Because these ducts
are not involved in active reabsorption, clear cytoplasm and the absence of the
brush border are characteristic of them. (200×)

Collecting ducts
(ducts of Bellini)

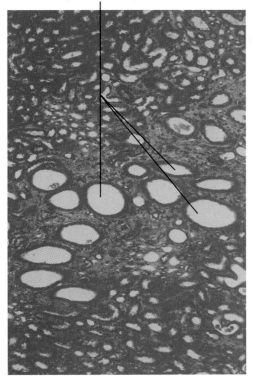

FIGURE 17.16
LM of a section through the renal medulla. Shown
is a large lumen of collecting ducts (ducts of
Bellini or papillary ducts), which are formed by
the fusion of collecting tubules and are lined by
simple columnar cells. The ducts are involved in
some absorption of water. (100×)

Lumen   Transitional   Connective
epithelium   tissue (lamina
propria)

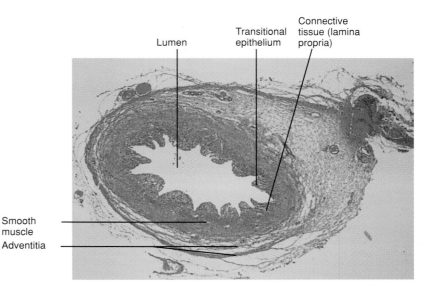

Smooth
muscle
Adventitia

FIGURE 17.17
LM of a cross section of a fetal ureter at low magnification. The lumen
appears narrow owing to the projection of transitional epithelium and
underlying connective tissue. The smooth muscles of the muscularis
form inner and outer longitudinal bundles of fibers. Sandwiched
between these two muscle layers is the circular muscle layer. The
adventitia forms the outermost covering of the ureter. (40×)

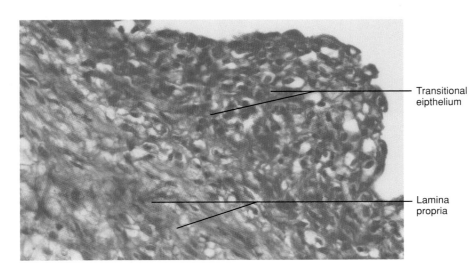

Transitional
eipthelium

Lamina
propria

FIGURE 17.18
LM of a fetal ureter in cross section, illustrating the morphology of transitional
epithelium. Below the epithelium lies the connective tissue of the lamina propria.
(400×)

FIGURE 17.19

LM of a cross section through an adult ureter. The lumen is surrounded by folded transitional epithelium. Below the epithelium lies the connective tissue of the lamina propria. An inner longitudinal muscular layer is surrounded by a circular layer of smooth muscle fibers. An outer longitudinal layer surrounds the circular layer. The loose connective tissue of the adventitia contains blood vessels and surrounds the muscularis. (40×)

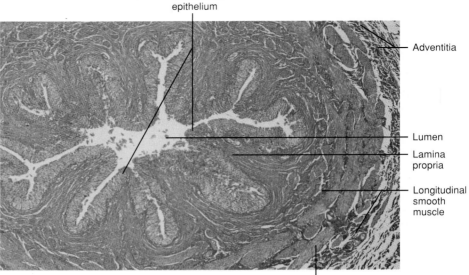

Transitional epithelium

Adventitia

Lumen

Lamina propria

Longitudinal smooth muscle

Circular smooth muscle

FIGURE 17.20

LM of a cross section of the adult ureter mucosa. The epithelium is a thick transitional layer in a relaxed state. The underlying connective tissue is the lamina propria. An inner longitudinal layer of smooth muscle fibers, and a middle circular layer of smooth muscle fibers, can be identified in the micrograph. (200×)

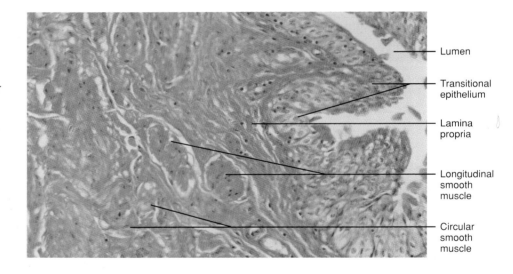

Lumen

Transitional epithelium

Lamina propria

Longitudinal smooth muscle

Circular smooth muscle

FIGURE 17.21

LM of an adult urinary bladder in a relaxed state. The body wall layers are similar to the ureter except for their thickness. The micrograph shows that the mucosa is highly folded and lined by a thick transitional epithelium, and the lamina propria connective tissue is wider than that in the ureter. Inner longitudinal and middle circular muscle fibers can also be seen in the micrograph. (100×)

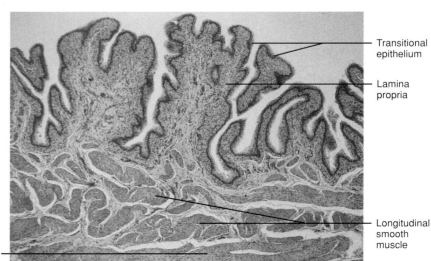

Transitional epithelium

Lamina propria

Longitudinal smooth muscle

Circular smooth muscle

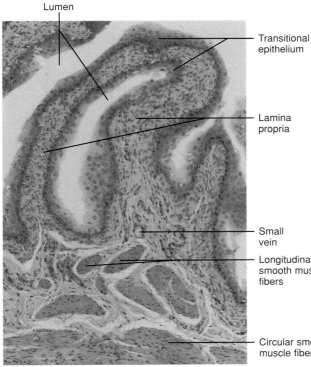

Lumen

Transitional epithelium

Lamina propria

Small vein

Longitudinal smooth muscle fibers

Circular smooth muscle fibers

FIGURE 17.22
LM of a cross section through the mucosa and part of the muscularis of the urinary bladder. The mucosa is extensively folded and is lined by a transitional epithelium. Below the epithelium lies the lamina propria of loose connective tissue. Sandwiched between the bundles of longitudinal smooth muscle fibers are circular muscle fibers. (200×)

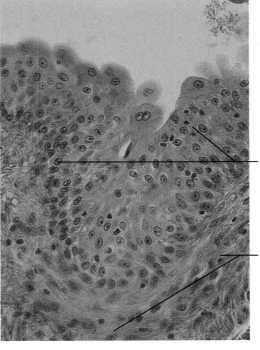

Transitional eipthelium

Lamina propria

FIGURE 17.23
LM of a section through the bladder mucosa. The epithelium is highly stratified in a relaxed state, as shown in this micrograph. Lamina propria connective tissue underlies the epithelium. (200×)

FIGURE 17.24
LM of a penile urethra in cross section. The penile urethra (cavernous urethra) passes through the corpus cavernosum urethrae (corpus spongiosum). The epithelial lining of the urethra is columnar in the prostate region but becomes pseudostratified or stratified columnar with patches of squamous epithelium in the rest of the urethra. Underlying the epithelium is the loose fibroelastic lamina propria. Blood vessels and surrounding corpus spongiosum can also be seen in the micrograph. (100×)

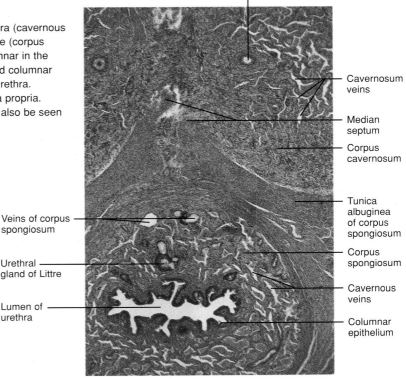

Deep artery of corpus cavernosum penis

Cavernosum veins

Median septum

Corpus cavernosum

Tunica albuginea of corpus spongiosum

Corpus spongiosum

Cavernous veins

Columnar epithelium

Veins of corpus spongiosum

Urethral gland of Littre

Lumen of urethra

FIGURE 17.25
LM of epithelium and lamina propria of a penile urethra in cross section. Stratified columnar epithelium lines this part of the urethra. Lamina propria of fibroelastic connective tissue can be seen below the epithelium. (200×)

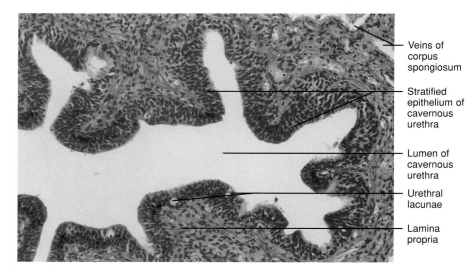

Veins of corpus spongiosum

Stratified epithelium of cavernous urethra

Lumen of cavernous urethra

Urethral lacunae

Lamina propria

FIGURE 17.26
LM of prostatic penile urethra in cross section. As shown in the micrograph, the epithelial lining in this part of the urethra is stratified columnar. The underlying loose fibroelastic connective tissue of the lamina propria can also be identified. (400×)

Stratified columnar epithelium

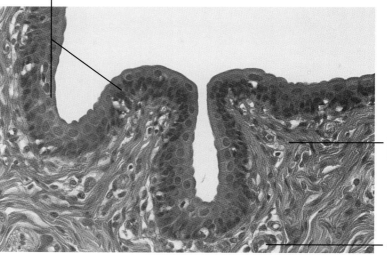

Lamina propria

Arteriole

Paraurethral glands

Lamina propria

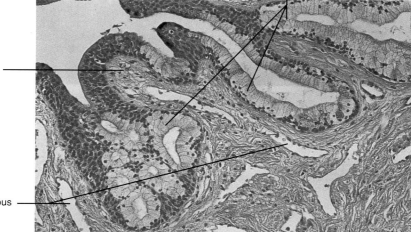

Cavernous veins

FIGURE 17.27
LM of the terminal penile urethra (fossa navicularis) in cross section. Visible is the stratified columnar epithelium with a bordering of the paraurethral glands. The lamina propria, with fibroelastic connective tissue, underlies the epithelium. Large concentrations of blood vessels can also be seen in the micrograph. (200×)

Endocrine System

The **endocrine system** includes the ductless glands that secrete their products (hormones) directly into the blood or lymph system. The endocrine glands in general are separate entities: the **pineal, pituitary, thyroid, parathyroid** and **adrenal.** However, scattered masses of endocrine cells may be present in the tissue of the exocrine glands: e.g., the **islets of Langerhans** in the **pancreas,** the **Leydig cells** in the testis interstitium, and the **corpora lutea** in the **ovaries.** These combinations of organs and glands are classified as **mixed glands.**

The endocrine glands are simple glands with glandular cells surrounded by connective tissue and an elaborate system of **fenestrated** blood capillaries that course through the endocrine tissue. Embryologically, the endocrine glands are derivatives of all three germinal layers:

1. The **pituitary** or **hypophysis,** the **adrenal medulla,** and the **chromaffin bodies** are of **ectodermal** origin.

2. The **testes,** the **ovaries,** and the **adrenal cortex** are derivatives of the **mesoderm.**

3. The **parathyroid,** the **thyroid,** and the **islets of Langerhans** are derived from the **endoderm.**

Chemically, the hormones may be **cholesterol** derivatives (**steroids**), **amino acids, proteins, glycoproteins,** or **peptides.** Because some of the hormones are lipid derivatives, they can diffuse through the cell membrane into the cell and bind with their receptor molecules in the cytoplasm or the nucleus. Hormones that are not lipid soluble bind to their respective receptor molecules on the membrane of target cells and bring about their hormonal action. A good example of such an interaction is the binding of a protein or peptide hormone (**first messenger**) to its receptor on the membrane, activating **adenylate cyclase,** an enzyme that converts **adenosine triphosphate (ATP)** into **cyclic adenosine monophosphate (cAMP).** The cAMP acts as a second messenger in the cytoplasm and mediates hormonal action in the cell.

Endocrine Glands—The Pituitary (Hypophysis)

The pituitary gland is attached to the median eminence of the **tuber cinereum** of the **hypothalamus** by an **infundibular stalk.** A bony depression, the **sella** turcica of the **sphenoid bone,** and a fibrous capsule, **diaphragma sellae,** protects the lateral and inferior surfaces of the pituitary. Morphologically, the pituitary can be divided into four parts:

1. The **pars anterior** or **pars distalis,** which secretes several hormones: **human growth hormone (GH)** or **somatotropin (STH),** a protein; **prolactin (PRL)** or **lactogenic hormone (LTH),** a protein; **thyroid stimulating hormone (TSH)** or **thyrotrophin,** a glycoprotein; **gonadotropin** or **follicle stimulating hormone (FSH)** and **luteinizing hormone (LH),** a glycoprotein; **adrenocorticotrophic hormone** or **corticotrophin (ACTH),** a polypeptide; and **melanocyte stimulating hormone (MSH),** a peptide.

2. The **infundibular stalk** of the pituitary includes the **pars tuberalis.** The cells in this region have granular cytoplasm and glycogen and are closely associated with blood vessels. The function of these cells is not known.

3. The **pars intermedia,** a poorly developed lobe, lies between the lobes of the anterior and posterior pituitary. It accounts for 2% of the pituitary. In adults, the pars intermedia cells blend into the **pars distalis.**

4. The **neurohypophysis** or **pars nervosa (pars posterior)** includes the **infundibular stem,** the median eminence of the **tuber cinereum,** and the infundibular process. All three divisions have the same cell types and share the same nerve and blood supply. **Supraoptic** and **paraventricular** nuclei in the hypothalamus send approximately 100,000 unmyelinated fibers (**hypothalamic-hypophyseal track**) to the neurohypophysis. The cell bodies of the supraoptic nucleus elaborate the **antidiuretic hormone (ADH) vasopressin,** whereas the **paraventricular** cell bodies secrete **oxytocin.** Both hormones are released by the terminals of the hypothalamic neurons as needed.

The Thyroid Gland

The bilobed thyroid gland is anteriorly located in the neck. It plays an important endocrine function in regulating the **basal metabolic rate (BMR)** of all cells in the body. The two lobes of the thyroid are connected by a narrow **isthmus.** The lobes of the

thyroid are divided into lobules by the continuation of the connective tissue from the surface capsule that envelops the thyroid. Each lobule is divided into follicles, the structural units of the gland. Each follicle is composed of simple cuboidal cells forming spheroidal structures. The follicles enclose a jelly-like colloidal substance. The colloid is a mixture of **thyroglobulin,** a glycoprotein containing iodinated amino acids and enzymes.

Also present in the thyroid are **parafollicular cells (C cells).** These cells are larger than follicular cells and have a nucleus placed eccentrically. They stain lightly and form small clusters between the follicles. Parafollicular cells secrete **thyrocalcitonin (or calcitonin).**

The thyroid gland secretes hormones that regulate the metabolic rate of all cells in the body. **Thyroxine** from the thyroid increases cellular metabolism. **Thyrocalcitonin (calcitonin),** from the parafollicular cells, lowers plasma calcium and increases bone formation.

The Parathyroid Glands

Generally, there are two pairs of small, oval-shaped parathyroid glands that lie on the posterior surface of the thyroid gland, two on each lobe. Each gland is enclosed by a thin connective tissue capsule separating it from the thyroid gland. The parenchyma of the gland is composed of cords of epithelial cells, the **chief cells (principal cells),** and **oxyphil cells.** Chief cells, more prevalent than oxyphil cells, can be identified by their large vesicular nuclei and by the few granules in the clear cytoplasm. The chief cells of the parathyroid glands secrete a hormone with a single chain protein, **parathormone (parathyroid hormone),** which stimulates the **osteoclast** cells in the bone, resulting in greater bone resorption and increased blood calcium levels. Parathormone also decreases blood **phosphate** levels.

Oxyphil cells are present in small or large groups, are larger than chief cells, and have dark-staining cytoplasm and small nuclei. In humans, oxyphil cells increase in number after 5 to 7 years and maximize in number after puberty. The function of the oxyphil cells in the parathyroid glands is not clear.

The Adrenal Glands (Suprarenal Glands)

The adrenal glands are pyramidal in shape and are surrounded by a fibrous capsule. The individual glands are located at each cranial pole of the kidneys. In cross section, two distinct regions can be identified: an outer **cortex** and an inner **medulla.** The cortex is a derivative of **mesodermal mesothelium,** and the medulla is derived from **ectodermal neural crest cells** and **autonomic ganglionic** tissue. The cortex of the adrenal gland can be further divided into three poorly defined zones:

1. The **zona glomerulosa** consists of columnar or pyramidal cells with densely staining spherical nuclei and cytoplasm. Lipid droplets can also be seen in the cytoplasm.

2. The **zona fasciculata** is the largest of the three zones. It can be identified by the long cords of irregular cuboidal or polyhedral cells with a centrally placed nucleus and cytoplasm containing lipid droplets.

3. The **zona reticularis** is the innermost zone of the adrenal cortex. The cell cords are arranged anastomostically. The nuclei are small and the cytoplasm has fewer lipid droplets in comparison with the zona fasciculata cells. Lipofuscin granules can also be found in these cells.

The reticularis of the cortex blends into the centrally located **medulla.** The cells in the medulla range from polyhedral to ovoid in shape and have large nuclei. They are organized in clusters and cords and are surrounded by arterioles, venules, and capillaries. The cells of the medulla are called **chromaffin** (or **pheochrome**) because they stain brown when oxidized by potassium bicarbonate. These cells secrete **catecholamines (epinephrine, norepinephrine, and dopamine).**

The adrenal cortex is essential for survival. The **zona glomerulosa** secretes **mineralocorticoids (aldosterone** and **deoxycorticosterone),** which are important for water and electrolyte balance. The **zona fasciculata** secretes **glucocorticoid hormones (cortisone** and **hydrocortisone).** The **zona reticularis** secretes female sex hormones (**estrogen** and **progesterone**) and the male hormones or **androgens (dehydroepiandrosterone).**

The Pineal Gland or Epiphysis Cerebri

This endocrine gland is a small, cone-shaped structure attached by a short stalk to the roof of the **third ventricle** of the brain. The pineal gland is surrounded by a **pia mater** capsule. The main body of the gland forms lobules that are separated by connective tissue septa. **Neuroglia (astrocytes** and **microglia)** and **epithelioid** cells or **pinealocytes** are the main cells of the lobules. Retrogressive changes in the pineal gland, which begin after puberty, lead to increased connective tissue and **corpora aranacea (brain sand)** in later stages of life. Functionally, the pineal is an endocrine gland that secretes **melatonin** and possibly other substances. Melatonin secretion in humans may be linked to delayed sexual development.

The Endocrine Pancreas

The pancreas serves as both an exocrine gland (secreting hydrolytic enzymes and bicarbonate) and an endocrine gland. The endocrine cells in the pancreas are localized in small clusters, the **islets of Langerhans.** There are three distinct populations of endocrine cells in the islets: the **alpha cells,** which secrete **glucagon;** the **beta cells,** which secrete **insulin;** and the **delta cells,** which secrete **somatostatin.** Insulin and glucagon regulate glucose levels in the blood. Somatostatin has a wide variety of functions associated with the gastrointestinal tract and the regulation of the secretion of insulin, glucagon, and human growth hormone.

Endocrine System

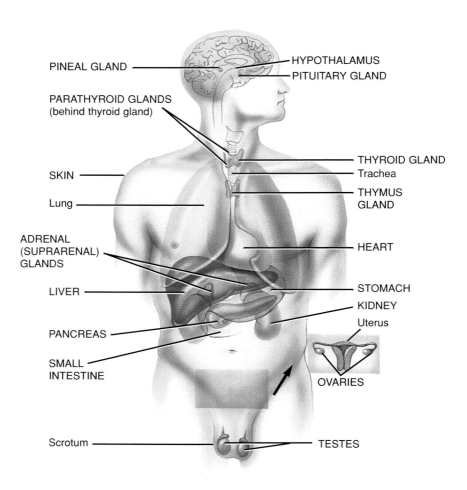

PINEAL GLAND

HYPOTHALAMUS

PITUITARY GLAND

PARATHYROID GLANDS
(behind thyroid gland)

THYROID GLAND

SKIN

Trachea

Lung

THYMUS
GLAND

ADRENAL
(SUPRARENAL)
GLANDS

HEART

LIVER

STOMACH

KIDNEY

PANCREAS

Uterus

SMALL
INTESTINE

OVARIES

Scrotum

TESTES

FIGURE 18.1
Diagram of the locations of the endocrine glands, organs containing endocrine tissue,
and associated structures.

Pituitary

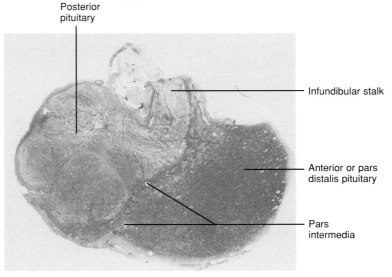

Posterior pituitary

Infundibular stalk

Anterior or pars distalis pituitary

Pars intermedia

FIGURE 18.2
Light micrograph (LM) of the pituitary (hypophysis), reflecting different regions: the anterior pituitary (pars distalis or adenohypophysis), the pars intermedia, the posterior pituitary (neurohypophysis), and the infundibular stalk. (1×)

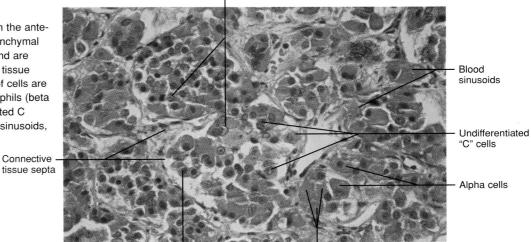

Chromophobes

Blood sinusoids

Undifferentiated "C" cells

Alpha cells

Connective tissue septa

Basophil cells

Sinusoids

FIGURE 18.3
LM of a cross section through the anterior lobe of the pituitary. Parenchymal cells are arranged in cords and are separated by thin connective tissue septa. Within this collection of cells are acidophils (alpha cells), basophils (beta and delta cells), undifferentiated C cells (chromophobes), blood sinusoids, and capillaries. (400×)

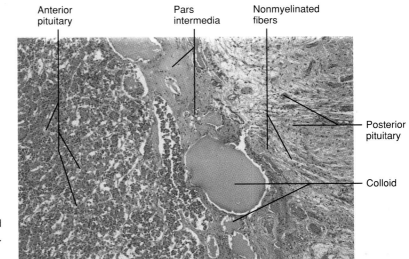

Anterior pituitary

Pars intermedia

Nonmyelinated fibers

Posterior pituitary

Colloid

FIGURE 18.4
LM of a sagittal section through the anterior pituitary, the pars intermedia, and the posterior pituitary. The anterior pituitary lobe stains darker because of massive numbers of nucleated cells. The pars intermedia is part of the anterior pituitary, both of which have the same embryological origin. The cells in the pars intermedia are basophilic. Eosinophilic colloid fills small cavities. The eosinophilic cells secrete proopiocortin (similar to the precursor peptide found in corticotrophs), which later becomes two forms of melanocyte stimulating hormone (MSH): endorphins and lipotropins. MSH stimulates the melanocytes in the skin to secrete melanin, a skin pigment. The posterior pituitary is the lightly stained area in the micrograph. It contains nonmyelinated fibers. The cell bodies of the fibers are located in the hypothalamus. (200×)

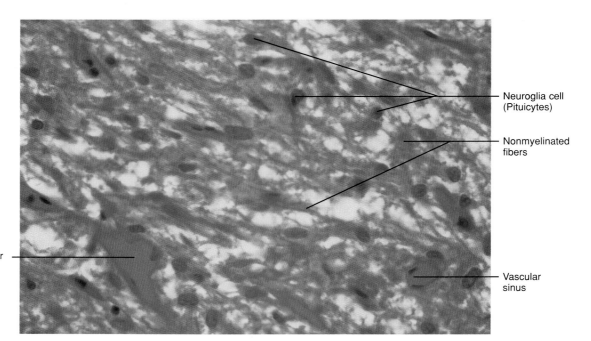

Vascular sinus

Neuroglia cell (Pituicytes)

Nonmyelinated fibers

Vascular sinus

FIGURE 18.5

LM of a cross section through the posterior pituitary (neurohypophysis). This area contains non-myelinated fibers of neurosecretory cells, the cell bodies of which are located in the hypothalamus. Also visible in the micrograph are neuroglia cells and fine blood capillaries within the fibers. (400×)

Thyroid and Parathyroid

FIGURE 18.6
(a) Diagram (anterior view) of the thyroid gland. (b) Diagram (posterior view) of the thyroid gland, indicating the locations of the parathyroid glands. Also shown are the supporting structures, blood vessels, and other organs in the area.

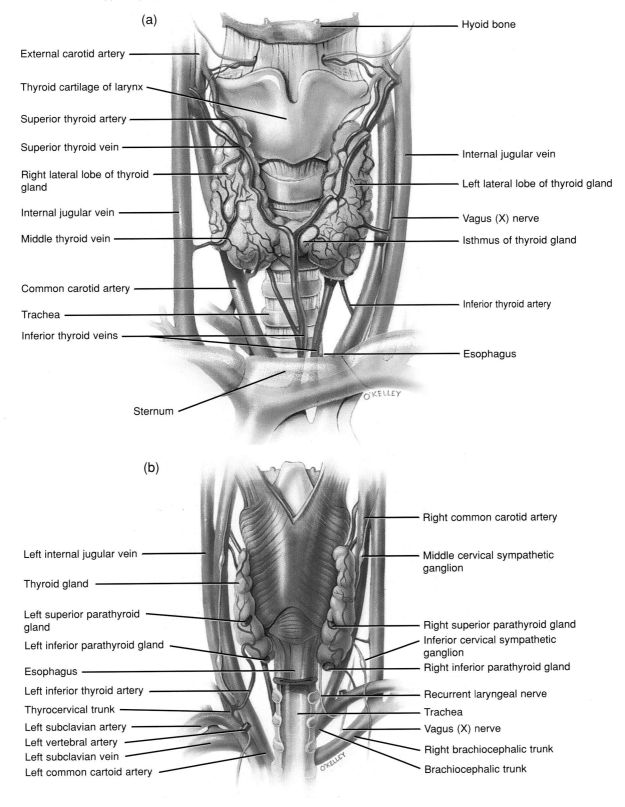

(a)

External carotid artery

Thyroid cartilage of larynx

Superior thyroid artery

Superior thyroid vein

Right lateral lobe of thyroid gland

Internal jugular vein

Middle thyroid vein

Common carotid artery

Trachea

Inferior thyroid veins

Sternum

Hyoid bone

Internal jugular vein

Left lateral lobe of thyroid gland

Vagus (X) nerve

Isthmus of thyroid gland

Inferior thyroid artery

Esophagus

(b)

Left internal jugular vein

Thyroid gland

Left superior parathyroid gland

Left inferior parathyroid gland

Esophagus

Left inferior thyroid artery

Thyrocervical trunk

Left subclavian artery

Left vertebral artery

Left subclavian vein

Left common cartoid artery

Right common carotid artery

Middle cervical sympathetic ganglion

Right superior parathyroid gland

Inferior cervical sympathetic ganglion

Right inferior parathyroid gland

Recurrent laryngeal nerve

Trachea

Vagus (X) nerve

Right brachiocephalic trunk

Brachiocephalic trunk

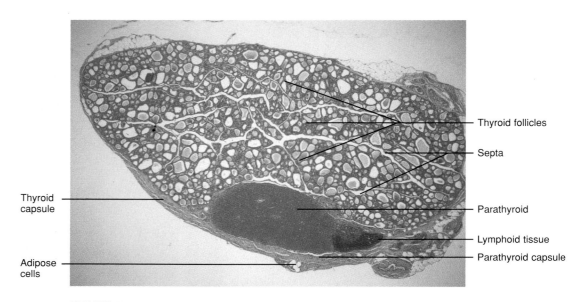

Thyroid follicles

Septa

Thyroid capsule

Parathyroid

Lymphoid tissue

Parathyroid capsule

Adipose cells

FIGURE 18.7

LM of a section of the thyroid, the parathyroid, and a small lymphoid nodule. The thyroid gland is divided into lobules separated by connective tissue. A thin capsule surrounds the thyroid and parathyroid glands. Within the thyroid are a large number of follicles that are functional units of the gland. (40×)

(a)

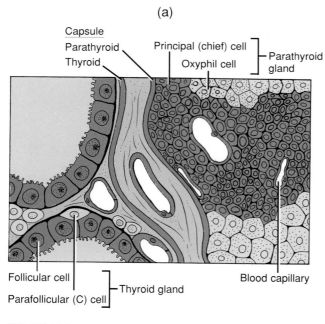

Capsule
Parathyroid
Thyroid

Principal (chief) cell
Oxyphil cell

Parathyroid gland

Follicular cell
Parafollicular (C) cell

Thyroid gland

Blood capillary

(b)

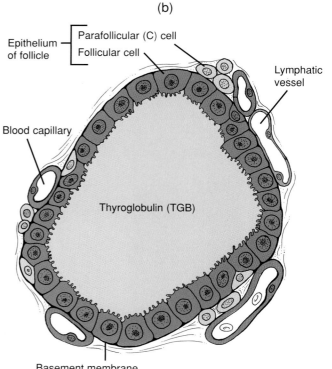

Epithelium of follicle

Parafollicular (C) cell
Follicular cell

Lymphatic vessel

Blood capillary

Thyroglobulin (TGB)

Basement membrane

FIGURE 18.8

(a) Diagram of a portion of the thyroid gland (left), a portion of the parathyroid gland (right), and cells associated with the glands. (b) Diagram of a single thyroid follicle, including follicular and parafollicular cells, blood capillaries, and supporting structures associated with the thyroid follicle.

FIGURE 18.9
LM of thyroid follicles in cross section. The follicles are separated by thin connective tissue and lined by cuboidal follicular cells. Some of the follicles contain colloid that appears almost homogeneous. Fairly large, lightly stained parafollicular cells can be seen between the follicles. (200×)

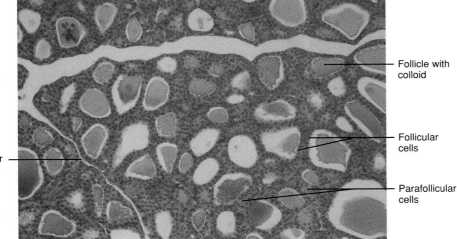

Follicle with colloid

Follicular cells

Parafollicular cells

Interlobular connective tissue

FIGURE 18.10
LM of thyroid follicles in cross section. The follicular epithelial cells in a follicle are cuboidal in shape and are fairly small compared with the larger parafollicular cells (C cells), which have lightly staining cytoplasm. Also, the parafollicular cells are found at the periphery of the thyroid follicle. Some of the thyroid follicles are filled with colloid, a mixture of triiodothyranine (T3) and tetraiodothyranine (T4). (400×)

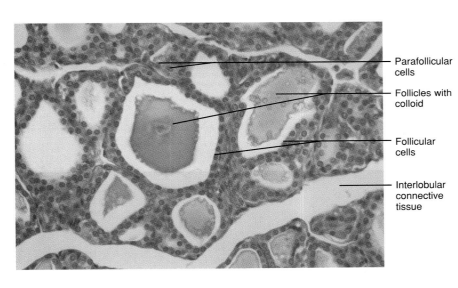

Parafollicular cells

Follicles with colloid

Follicular cells

Interlobular connective tissue

Retracted colloid

FIGURE 18.11
LM of the enlarged interfollicular space of a thyroid gland. Parafollicular cells (C cells or clear cells) can be identified by their large size and clear, lightly staining cytoplasm. Parafollicular cells synthesize calcitonin, a hormone. (1000×)

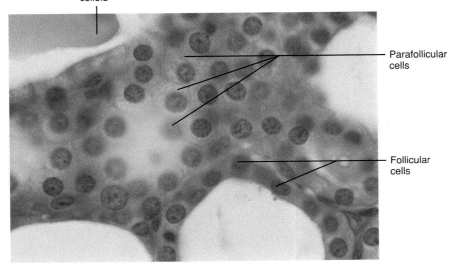

Parafollicular cells

Follicular cells

FIGURE 18.12
LM of a midsagittal section of an isolated parathyroid gland. The gland is surrounded by a thin connective tissue capsule. The connective tissue from the capsule enters the gland and forms delicate septa, dividing the parathyroid parenchyma into masses of cord-like secretory cells. Blood vessels, sinusoids, and colloid concentrations can be seen in the micrograph. (40×)

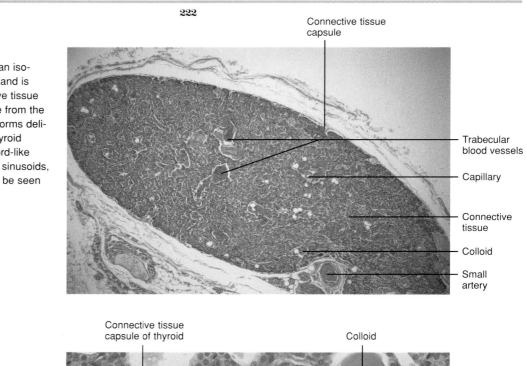

Connective tissue capsule

Trabecular blood vessels

Capillary

Connective tissue

Colloid

Small artery

FIGURE 18.13
LM of a cross section of the parathyroid and thyroid glands. As can be seen in the micrograph, the two glands are separated from each other by their respective connective tissue capsules. The parenchyma of the parathyroid gland displays chief (principal) cells and oxyphil cells. The arrangement of cells in the thyroid follicles differs from that in the parathyroid. (400×)

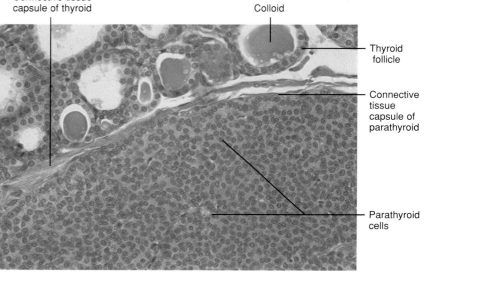

Connective tissue capsule of thyroid

Colloid

Thyroid follicle

Connective tissue capsule of parathyroid

Parathyroid cells

FIGURE 18.14
LM of the parathyroid parenchyma, at a higher magnification. The most abundant chief (principal) cells can be identified by the prominent nuclei and scant cytoplasm. These cells secrete parathormone, the parathyroid hormone. The oxyphil cells in the parenchyma are larger and fewer in number, have poorly staining nuclei, and have strongly eosinophilic (oxyphilic) cytoplasms. Generally, these cells are found in clumps. Their function is unknown. (1000×)

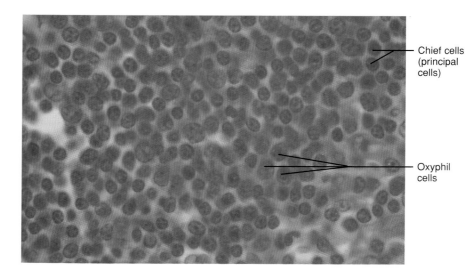

Chief cells (principal cells)

Oxyphil cells

Adrenal or Suprarenal Gland

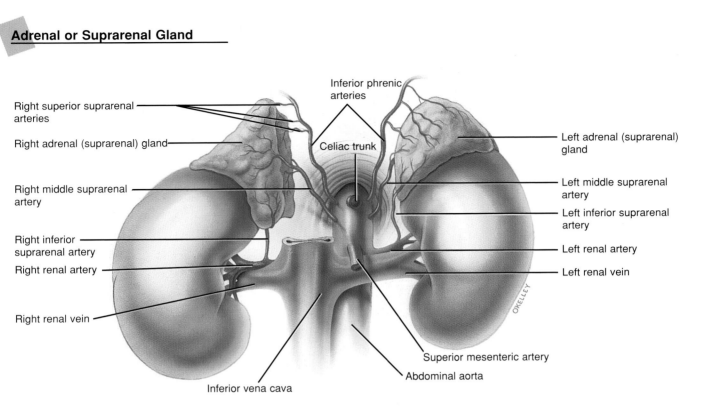

Right superior suprarenal arteries

Right adrenal (suprarenal) gland

Right middle suprarenal artery

Right inferior suprarenal artery

Right renal artery

Right renal vein

Inferior phrenic arteries

Celiac trunk

Left adrenal (suprarenal) gland

Left middle suprarenal artery

Left inferior suprarenal artery

Left renal artery

Left renal vein

OKELLEY

Superior mesenteric artery

Abdominal aorta

Inferior vena cava

FIGURE 18.15
Diagram of the kidneys and the locations of the adrenal glands (suprarenal glands). Also shown in the diagram are blood vessels and other supporting structures in the area.

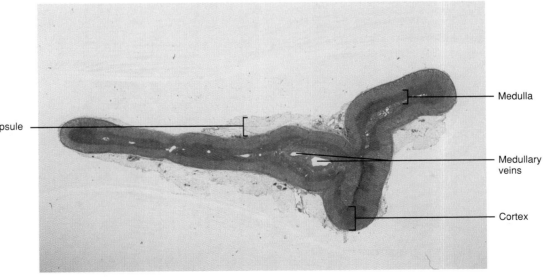

Medulla

Capsule

Medullary veins

Cortex

FIGURE 18.16
LM of the midsagittal section of the adrenal (suprarenal) gland. The cortex and medulla regions can be easily identified. Large medullary veins form the characteristic identifying feature in the medulla of the adrenal gland. A connective tissue capsule surrounds the suprarenal gland. (1×)

FIGURE 18.17
LM of a section through the adrenal gland. A connective tissue capsule with blood vessels, and the cortex with three identifiable zones (zona glomerulosa, zona fasciculata, and zona reticularis), can be seen in the micrograph. Also identifiable are the medulla region and the lumen of the medullary vein. (100×)

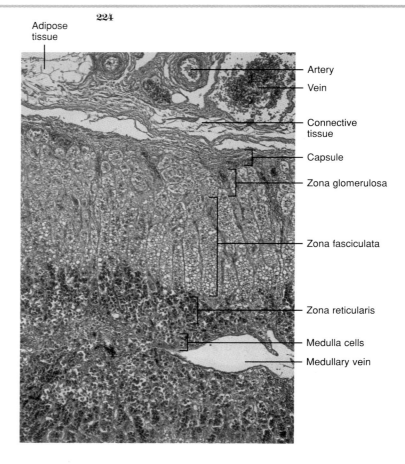

Adipose tissue

Artery

Vein

Connective tissue

Capsule

Zona glomerulosa

Zona fasciculata

Zona reticularis

Medulla cells

Medullary vein

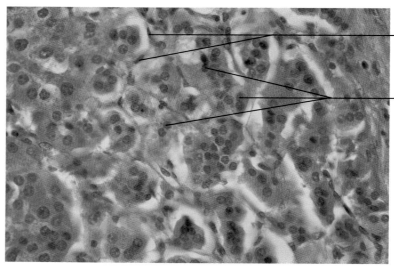

Sinusoidal capillaries and endothelial cells

Cells of the zona glomerulosa

FIGURE 18.18
LM of the zona glomerulosa of the adrenal gland in cross section, at a higher magnification. The cells in this zone are small, are ovoid in shape, and form small clusters that are separated by connective tissue. The secretory cell nuclei stain dark and are surrounded by small amounts of cytoplasm. The zona glomerulosa forms a narrow band of cells just below the adrenal capsule. The cells of the glomerulosa secrete the mineralocorticoid hormone, aldosterone. (400X)

Capillary
sinusoids

Anastomosing columns
of zona reticularis cells

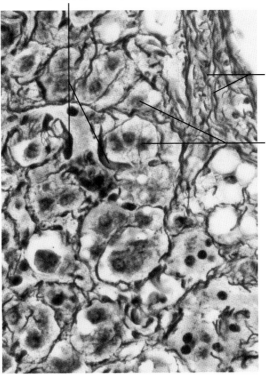

Connective
tissue

Cells of
the zona
fasciculata

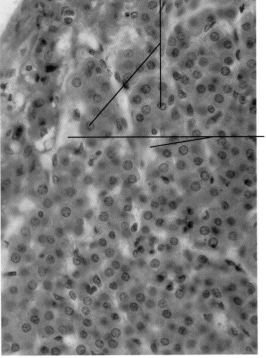

Capillary
sinusoids

FIGURE 18.19
LM of the zona fasciculata, at a higher magnification. The
cells in this zone are large and form cords or clusters of
secretory cells. The cells are separated by connective tissue
mixed with capillary sinusoids. The cell cytoplasm stains
poorly and is vacuolated, which indicates the presence of
lipids in the cell. The cells secrete glucocorticoid hormones,
primarily cortisol. (400×)

FIGURE 18.20
LM of the zona reticularis in cross section, at a higher magnification. The reticularis forms the innermost zone of the adrenal
cortex. The cells in this zone are compact, stain intensely, and
form branching cords that are separated by connective tissue
and wide capillary sinusoids. The secretory cells are smaller
than in the adjoining zona fasciculata. Androgen hormones,
mainly testosterone, are secreted by cells in this zone. (400×)

FIGURE 18.21
LM of the adrenal medulla, at a higher
magnification. The medulla cells are
known as chromaffin cells because they
stain brown when treated with chromium
salts. The cells in this region form compact clumps separated by fine strands of
connective tissue and capillary sinusoids.
Sinusoids drain blood into veins with
large lumen. The secretory cells of the
medulla have large granular nuclei with
intensely stained basophilic cytoplasms.
The adrenal medullary cells secrete the
catecholamine hormones norepinephrine,
epinephrine, and dopamine. (400×)

Cells of the
medulla

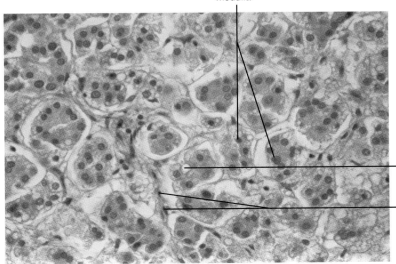

Capillary
sinusoids

Endothelial
cells of
sinusoids

Islets of Langerhans

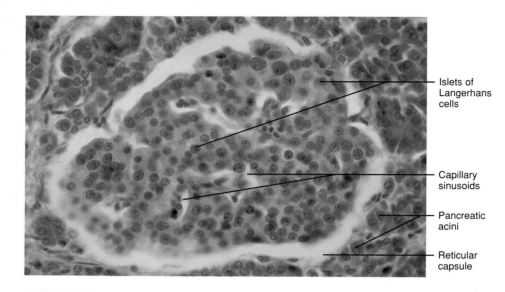

Islets of
Langerhans
cells

Capillary
sinusoids

Pancreatic
acini

Reticular
capsule

FIGURE 18.22

LM of a cross section through the pancreas, displaying the endocrine portion of the
gland, the islets of Langerhans. The gland is composed of clusters of secretory
cells supported by a connective tissue capsule and fenestrated capillaries. In the
micrograph, the islet is surrounded by the exocrine acini cells of the pancreas.
(400×)

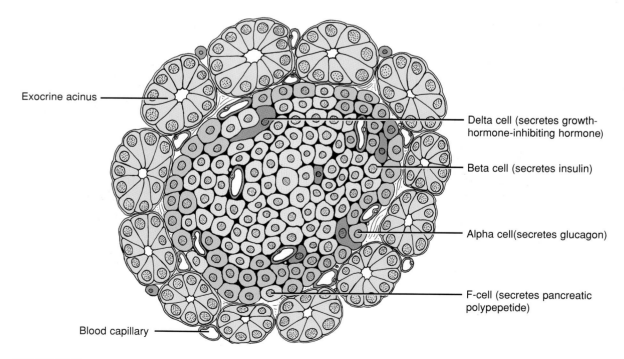

Exocrine acinus

Delta cell (secretes growth-
hormone-inhibiting hormone)

Beta cell (secretes insulin)

Alpha cell(secretes glucagon)

F-cell (secretes pancreatic
polypepetide)

Blood capillary

FIGURE 18.23

Diagram of a pancreatic islet (islet of Langerhans) and the surrounding acini. The
diagram illustrates the cell diversity among islet cells with respect to hormone
secretion.

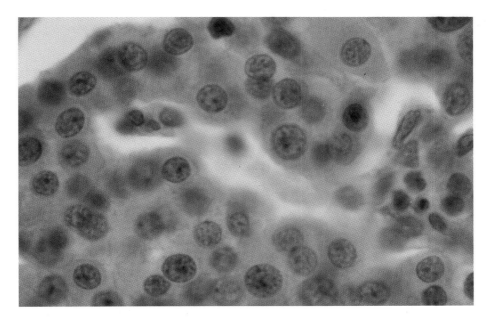

FIGURE 18.24
LM of a section of islet of Langerhans cells at a higher magnification. It is difficult to ascertain the type of cells in the islet without using special stain or immunofluorescent staining. However, three types of secretory cells are present in the islet: alpha cells, which secrete glucagon, beta cells, which secrete insulin, and delta cells, which secrete somatostatin. (1000×)

Pineal Gland (epiphysis cerebri)

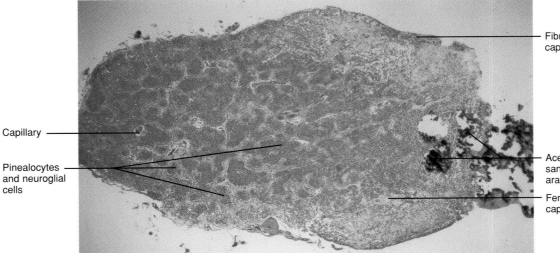

FIGURE 18.25
LM of the midsagittal section of the pineal gland (epiphysis cerebri) with its thin capsule of connective tissue (pia mater). The parenchyma of the gland is divided into lobules separated by connective tissue. At one end of the micrograph, purple lamellar bodies, the acervuli (brain sand or corpora aranacea), can be seen. The acervuli are localized in the capsule or in the septa between the lobules. (20×)

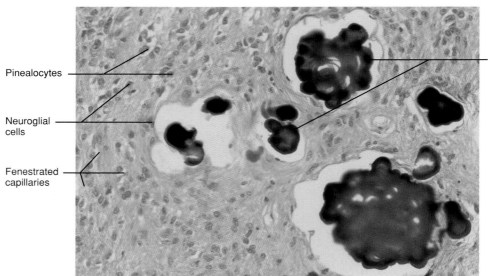

Pinealocytes

Neuroglial
cells

Fenestrated
capillaries

Acervuli (brain sand
or corpora aranacea)

FIGURE 18.26
LM of the pineal gland, showing large deposits of acervuli (brain sand or corpora
aranacea) in the parenchyma of the gland. Also visible in the micrograph are
pinealocytes (epithelioid cells) and neuroglial cells. (200×)

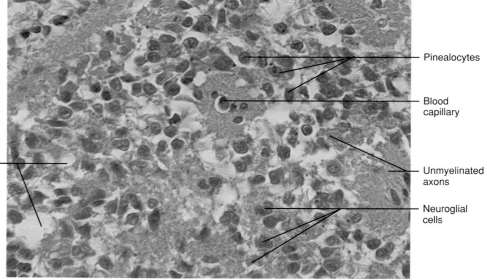

Fenestrated
capillaries
and sinuses

Pinealocytes

Blood
capillary

Unmyelinated
axons

Neuroglial
cells

FIGURE 18.27
LM of the pineal gland, at a higher magnification. Visible in the micrograph are the
pale-staining clumps of pinealocytes characterized by large ovoid nuclei. Neuroglial
cells and blood capillaries are also identifiable in the micrograph. (400×)

The Male Reproductive System

The **male reproductive system** consists of the **testes,** which produce **spermatozoa** and **androgens;** the glands that produce the fluid that facilitates the transfer of sperm; a series of ducts and passageways for the transport of sperm and fluids; and a copulatory organ, the **penis,** which delivers the spermatozoa to the female reproductive tract.

The organs and tubular passageways of the male reproductive system are: the **testes, seminiferous tubules,** the **rete testis,** the **ductus epididymis,** the **vas deferens,** the **ampulla of the ductus (vas) deferens,** the **seminal vesicles,** the **ejaculatory duct,** the **corpora cavernosum urethrae,** the **penile urethra,** the **prostate,** and the **bulbourethral glands.**

Testes **Spermatozoa** are produced through the process of spermatogenesis, which occurs in the testes. The testes are also the site of the production of androgens, the male sex hormones. Two testes are suspended in a testicular sac, the **scrotum.** The testes are surrounded by a **testicular capsule** composed of a **tunica vaginalis** (outer layer), a **tunica albuginea** (middle layer), and a **tunica vasculosa** (innermost layer). Internally, each testis is partitioned into approximately 250 pyramidal-shaped compartments, the **lobuli testis,** which are partially separated by connective tissue septa. Within each lobule are one to four extensively convoluted **seminiferous tubules.** The tubules are surrounded by a stroma of loose connective tissue. Several types of cells, including the endocrine interstitial **cells of Leydig,** are present in this tissue.

Seminiferous Tubules The seminiferous tubules are approximately 0.2 mm in diameter, 30 to 70 cm long, extensively convoluted, and covered by connective tissue called **peritubular tissue.** Each tubule is lined by a specialized stratified cuboidal seminiferous epithelium, **germinal epithelium.** The seminiferous epithelium consists of proliferating germinal cells. Here, spermatozoa are formed by the process of differentiation. Supporting or sustentacular **cells of Sertoli** are present in the interstitium. Sertoli cells are believed to play a role in the maturation of spermatozoa.

Rete Testis The straight tubular parts of the seminiferous tubules open into anastomosing channels, the rete testis. The rete is lined by spaced cuboidal or squamous cells with an occasional cuboidal cell with cilia. The epithelium rests on a basement membrane.

Ductuli Efferentes   The concentration of efferent ductules forms from 10 to 15 ductuli efferentes. These lobules emerge from the rete testis. The ductules are covered by connective tissue with underlying smooth muscle fibers. The epithelial lining is simple columnar supported by a basal lamina. Tall columnar cells of the epithelium are ciliated, whereas the short columnar cells are lined by microvilli. These cells absorb the excess fluid secreted by the seminiferous tubules.

Ductus Epididymis   The ductuli efferentes empty into a single, long, highly tortuous ductus epididymis. The epithelial lining of the ductus is pseudostratified and is composed of tall columnar cells and basal cells with pigment granules. The columnar cells are lined by **stereocilia** on their free surfaces. A basement lamina underlies the epithelium, and thin layers of smooth muscle form a circular layer.

Ductus Deferens (Vas Deferens)   The terminal, straight end of the ductus epididymis blends into the ductus deferens, which ascends from the scrotum, passes through the **inguinal region,** traverses retroperitoneally, descends along the side wall of the pelvis, and merges with the **urethra.** The epithelial lining is pseudostratified with tall columnar cells bearing stereocilia that rest on the basement lamina. The underlying lamina propria (with elastic fibers) forms longitudinal folds that give the mucosal surface a stellate appearance similar to that in a cross section of the vas deferens.

Ampulla of Ductus Deferens   At the terminal end, the ductus deferens forms a sac-like dilatation, the ampulla. The mucosa in this region forms folds that extend into the large lumen of the ampulla. The epithelium is simple columnar with secretory cells, and the body wall musculature is sparse.

Seminal Vesicles   Each seminal vesicle is an elongated evagination of the ductus deferens and is located posterior to the **prostate gland.** The epithelium is pseudostratified columnar with an underlying basement lamina. A highly vascularized **lamina propria** of loose connective tissue lies under the basement lamina. The epithelial secretory cells contain yellow granular pigment. The secretion is rich in **ascorbic acid, prostaglandins, globulin,** and **fructose,** which are essential for the maturation of **spermatozoa.**

Ejaculatory Duct   At the terminal segment of the genital duct is a short ejaculatory duct formed by the fusion of the ampulla and the excretory duct of the seminal vesicle. The ejaculatory duct opens into the urethra next to the **prostate utricle.** The epithelial lining is pseudostratified columnar or simple columnar.

Prostate   The **urethra,** at its origin from the **urinary bladder,** is surrounded by the prostate gland. Approximately 30 to 50 compound **tuboalveolar** glands make up the prostate. The prostate gland is surrounded by a highly vascularized **fibroelastic capsule** with smooth muscle cells. The mucosal epithelium ranges from simple columnar to pseudostratified columnar or low cuboidal with an underlying basement lamina and lamina propria. Prostate secretion is a thin, milky white fluid that is rich in acid phosphates, proteolytic enzymes, and fibrinolysin, which liquifies the semen. **Corpora amylacea,** if present in the prostate, are calcified prostatic secretory concentrations.

Bulbourethral Glands (Glands of Cowper)   The bulbourethral glands, or glands of Cowper, are small, paired tuboalveolar bodies in the connective tissue, behind the membranous urethra. The septa with connective tissue, smooth muscle, and elastic fiber divide the glands into lobules; the glands may be alveolar, saccular, or tubular in morphology. The bulbourethral glands secrete a clear viscid mucus that facilitates sperm movement and transportation.

Penis   The penis serves as a copulatory organ and a common outlet for seminal fluid and urine. The penis is composed of three cylinders of erectile tissue: two **corpora cavernosa penis** and a single **corpus cavernosum urethrae** or **corpus spongiosum.** The penile urethra is surrounded by corpus spongiosum. The three erectile tissues are surrounded by subcutaneous connective tissue that is rich in smooth muscle and elastic fibers. The thin overlying skin of the penis is firmly attached to the underlying connective tissue. Distally, the skin folds over the **glans penis** to form an inverted covering, the **prepuce.** Proximally, small sweat glands are present in the skin with occasional sebaceous glands that are associated with hair follicles.

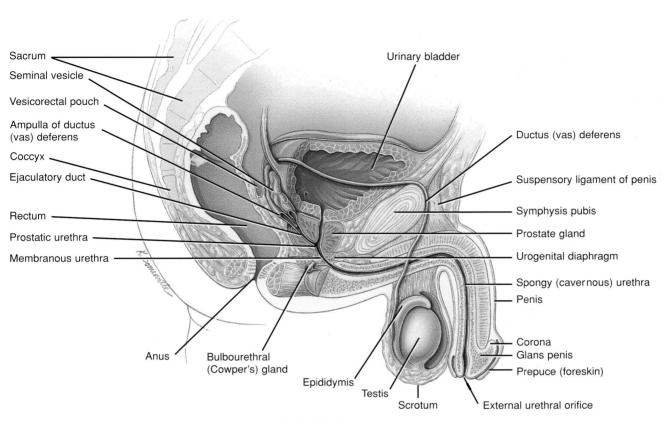

Sacrum

Seminal vesicle

Vesicorectal pouch

Ampulla of ductus
(vas) deferens

Coccyx

Ejaculatory duct

Rectum

Prostatic urethra

Membranous urethra

Urinary bladder

Ductus (vas) deferens

Suspensory ligament of penis

Symphysis pubis

Prostate gland

Urogenital diaphragm

Spongy (cavernous) urethra

Penis

Corona

Glans penis

Prepuce (foreskin)

Anus

Bulbourethral
(Cowper's) gland

Epididymis

Testis

Scrotum

External urethral orifice

Sagittal Section

FIGURE 19.1
Diagram of the male reproductive organs and supportive glands in a sagittal section. The diagram
shows supportive structures such as the scrotum, the membranous urethra, the urinary bladder,
the suspensory ligament of the penis, the symphysis pubis, and the urogenital diaphragm.

FIGURE 19.2
Diagram of a testis, showing the relationship
between the testis and the surrounding
tubular structures: the ductus (vas) deferens,
the seminiferous tubules, the rete testis, and
the ductus epididymis. Also shown in the
diagram are septa that separate the lobules
and the tunica albuginea.

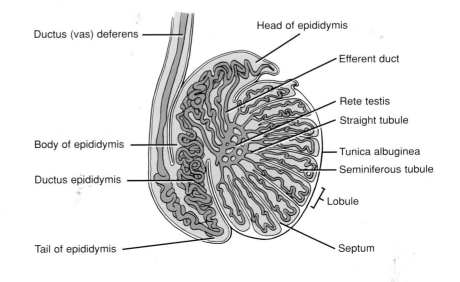

Ductus (vas) deferens

Head of epididymis

Efferent duct

Rete testis

Straight tubule

Tunica albuginea

Seminiferous tubule

Lobule

Body of epididymis

Ductus epididymis

Tail of epididymis

Septum

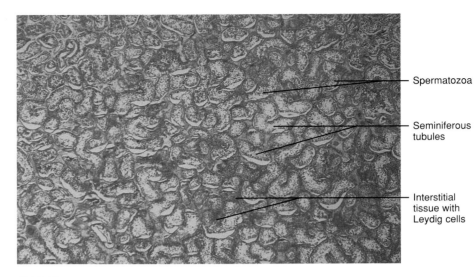

Spermatozoa

Seminiferous
tubules

Interstitial
tissue with
Leydig cells

FIGURE 19.3
Light micrograph (LM) of seminiferous tubules sectioned in several planes. The
micrograph illustrates the convoluted nature of the seminiferous tubules. When
mature, the tubules have a distinct lumen, and the epithelial cells produce spermatozoa by spermatogenesis. The interstitial spaces between the tubules consist of supporting connective tissue and endocrine Leydig cells. (20×)

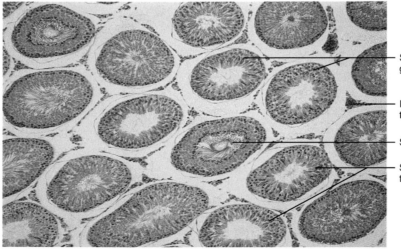

Sertoli and
germinal cells

Interstitial
tissue

Spermatozoa

Seminiferous
tubules

FIGURE 19.4
LM of the seminiferous tubules in cross section, at a higher magnification. The
micrograph illustrates the arrangement of Sertoli and germinal cells in the tubule
and the process of spermatogenesis. In some of the tubules, spermatozoa can
be seen in the lumen. (100×)

FIGURE 19.5
Diagram of germinal cells at different stages of cell division in the seminiferous tubules. The spermatogonium (2n) goes through cell division and, at successive stages of development, follows a sequence of primary spermatocyte (n), secondary spermatocyte (n), early spermatid (n), late spermatid (n), and finally maturing into spermatozoa (n). Also shown are supporting cells, interstitial endocrinocytes (cells of Leydig), and a sustentacular cell (Sertoli cell).

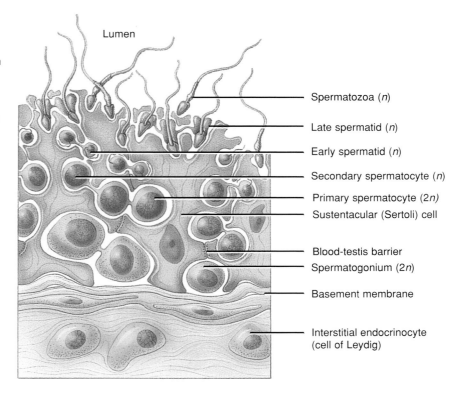

Lumen

Spermatozoa (n)

Late spermatid (n)

Early spermatid (n)

Secondary spermatocyte (n)

Primary spermatocyte (2n)

Sustentacular (Sertoli) cell

Blood-testis barrier

Spermatogonium (2n)

Basement membrane

Interstitial endocrinocyte (cell of Leydig)

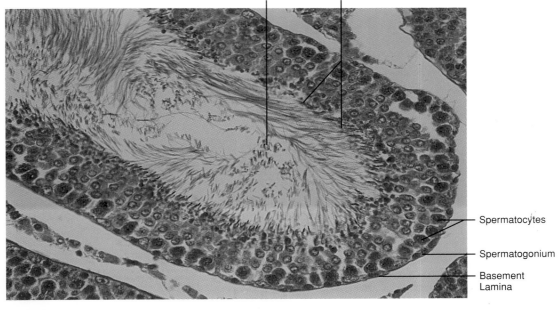

Spermatozoa

Secondary spermatocytes maturing on Sertoli cells

Spermatocytes

Spermatogonium

Basement Lamina

FIGURE 19.6
LM of a cross section through a seminiferous tubule. The micrograph illustrates different stages of spermatogenesis. The cells close to the basement lamina with spherical nuclei are the spermatogonia. Larger cells, also with spherical nuclei, are the spermatocytes. The spermatocytes divide and form into early and late spermatids, finally maturing into spermatozoa (n). The latter stages of development cannot be clearly differentiated in the micrograph. Mature spermatozoa can be seen in the lumen of the seminiferous tubule. (400×)

Spermatozoa

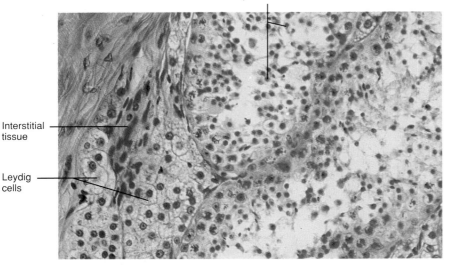

Interstitial
tissue

Leydig
cells

FIGURE 19.7
LM of a cross section through part of the
seminiferous tubules and the interstitial
tissue with concentrations of Leydig cells.
Also shown are spermatozoa at different
stages of development. The Leydig cells
secrete the principal male hormone, testosterone. (400×)

FIGURE 19.8
LM of Leydig cells and interstitial
tissue in cross section, at a higher
magnification. Surrounding the Leydig
cells is the connective tissue that
forms the margin of seminiferous
tubules. Also shown in the micrograph are spermatogonia germ cells
at different stages of development.
(1000×)

Spermatogonia

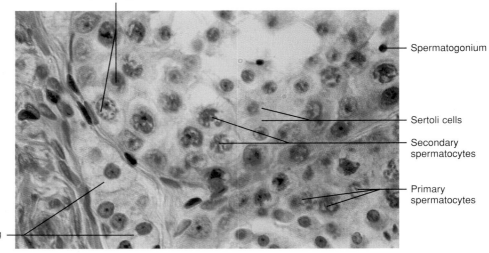

Spermatogonium

Sertoli cells

Secondary
spermatocytes

Primary
spermatocytes

Leydig
cells

FIGURE 19.9
LM of rete testis in cross section. Rete
testis is present in the mediastinum
testis. The rete testis is seminiferous
anastomosing, irregular tubules lined
by a single layer of low cuboidal or low
columnar cells. The lumen of the rete
widens before opening into the ductuli
efferentes (efferent ductules). (200×)

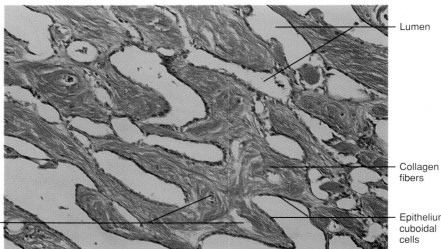

Lumen

Collagen
fibers

Epithelium
cuboidal
cells

Blood
vessels

FIGURE 19.10

LM of a cross section of the epididymis, possibly a storage place for the spermatozoa. The epididymis is a long, highly convoluted duct. This duct proceeds downward behind the testes toward the lower pole, and from there it becomes the ductus deferens (vas deferens). Shown in the micrograph are several cross sections through the same duct. This signifies the convoluted nature of the duct. The body wall of the duct displays pseudostratified epithelium with projections of stereocilia (microvilli) into the lumen. Also visible in the lumen are tufts of spermatozoa. In between the cross sections (longitudinal section in the upper and lower part of the micrograph) is the loose areolar connective tissue. (200×)

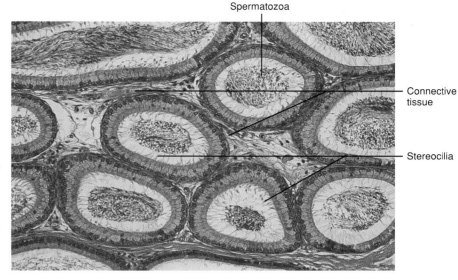

Spermatozoa — Connective tissue — Stereocilia

FIGURE 19.11

LM of the epididymis in cross section, at a higher magnification. Shown in the micrograph is the pseudostratified epithelial lining of the duct with stereocilia (microvilli) extending into the lumen. The stereocilia are believed to be involved in the absorption of excess fluid that accompanies the spermatozoa from the testes. Smooth muscle cells can be seen at the periphery of the body wall. Loose areolar connective tissue with fibroblast cells surrounds the epididymis. (400×)

Stereocilia — Spermatozoa — Nucleus — Pseudo-stratified columnar epithelium — Connective tissue — Lumen

FIGURE 19.12

LM of epididymis duct in cross section, at a still higher magnification. Shown in the micrograph are mature spermatozoa in the lumen of the epididymis. Also visible in the micrograph are the pseudostratified columnar epithelium with stereocilia and an underlying layer of smooth muscle and connective tissue. (1000×)

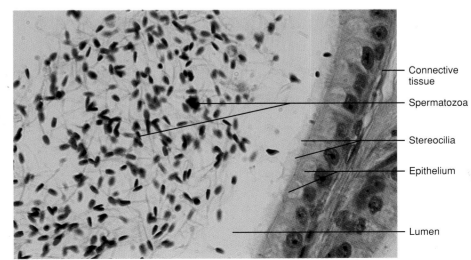

Connective tissue — Spermatozoa — Stereocilia — Epithelium — Lumen

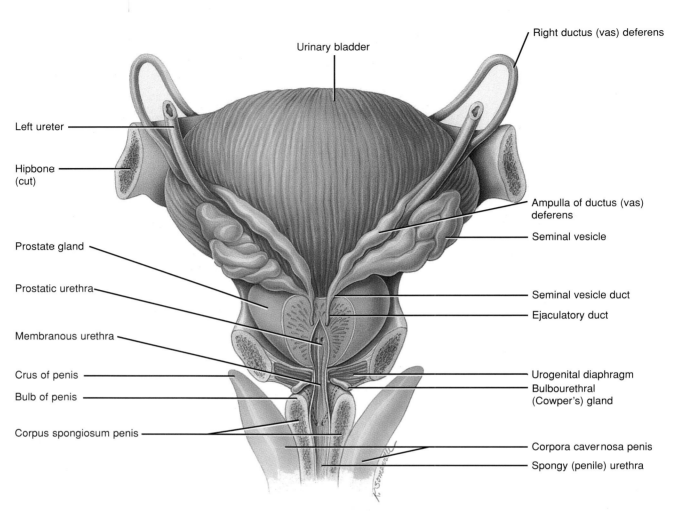

Urinary bladder

Right ductus (vas) deferens

Left ureter

Hipbone
(cut)

Ampulla of ductus (vas)
deferens

Seminal vesicle

Prostate gland

Prostatic urethra

Seminal vesicle duct

Ejaculatory duct

Membranous urethra

Crus of penis

Urogenital diaphragm

Bulbourethral
(Cowper's) gland

Bulb of penis

Corpus spongiosum penis

Corpora cavernosa penis

Spongy (penile) urethra

FIGURE 19.13
Diagram of male reproductive organs (posterior view) and their relationship
with surrounding supporting structures.

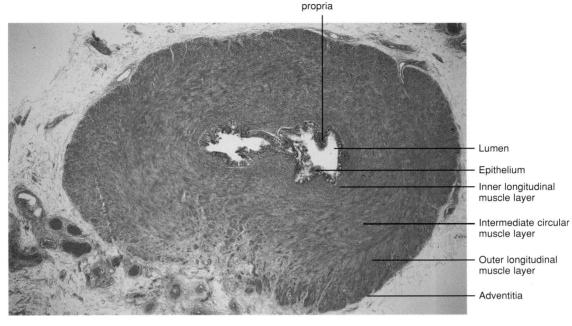

FIGURE 19.14

LM of a cross section of the ductus deferens (vas deferens) at a low magnification. The micro-graph shows the thick, muscular body wall of the ductus that facilitates the movement of the spermatozoa from the epididymis to the urethra. The smooth muscle fibers of the body wall are arranged in inner and outer longitudinal layers, and a circular middle layer of muscle fibers. The lumen is lined by pseudostratified columnar epithelium. Underlying the epithelium is the lamina propria. The epithelium and the lamina propria are thrown into folds that project into the lumen. Surrounding the outer muscle layer is a thin adventitia. (40×)

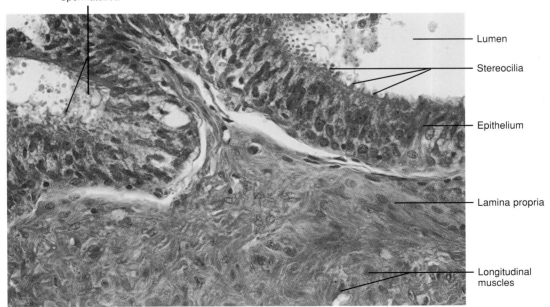

FIGURE 19.15

LM of a cross section of a portion of the body wall of the ductus deferens (vas deferens) at a higher magnification. The epithelial lining is of pseudostratified columnar cells surrounded by the lamina propria. Stereocilia (microvilli) can be seen extending from the epithelial cells toward the lumen. The stereocilia are believed to be absorptive in function. (200×)

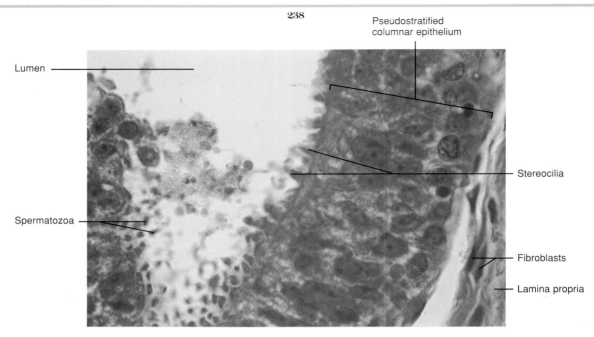

Lumen

Pseudostratified columnar epithelium

Stereocilia

Spermatozoa

Fibroblasts

Lamina propria

FIGURE 19.16

LM of a cross section through the body wall of the ductus deferens (vas deferens), displaying pseudostratified columnar epithelial lining and a small area of the lamina propria. Extending from the epithelium toward the lumen are stereocilia (microvilli). Fibroblast cells can also be identified in the micrograph. (1000×)

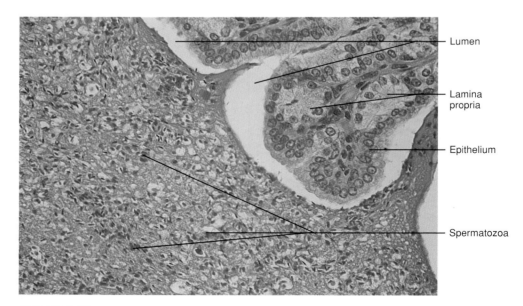

Lumen

Lamina propria

Epithelium

Spermatozoa

FIGURE 19.17

LM of a section through ampulla of ductus (vas) deferens. The micrograph illustrates the prominent mucosal folds lined by pseudostratified columnar epithelium. Underlying the epithelium is the lamina propria. The lumen shows large concentrations of spermatozoa mixed with secretory fluids. (400×)

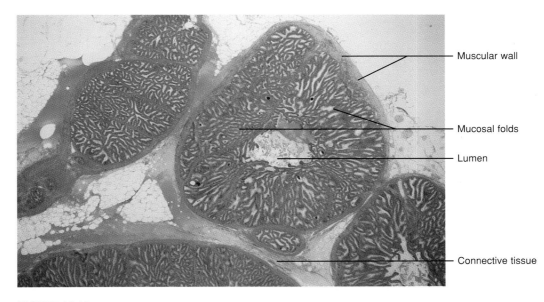

FIGURE 19.18

LM of a seminal vesicle in cross section, at low magnification. The seminal vesicle is a
convoluted evagination of the ductus deferens with an irregular lumen. The mucosa is
extensively folded, forming numerous primary folds that give rise to secondary and tertiary folds. The folds of the vesicle project into the lumen, where they further subdivide
and form into small compartments that give the lumen a honeycombed appearance. The
epithelial lining of the fold is pseudostratified columnar. The basal cells of the vesicle are
mixed cuboidal or simple columnar. The underlying lamina propria of loose connective
tissue forms a narrow band below the epithelium. Surrounding the lamina propria is a circular layer of smooth muscle fibers, which in turn is surrounded by thin layers of longitudinally arranged muscle fibers. The seminal vesicles secrete a viscid secretion that is rich
in glucose and prostaglandins. Spermatozoa are generally absent in the vesicles. (20×)

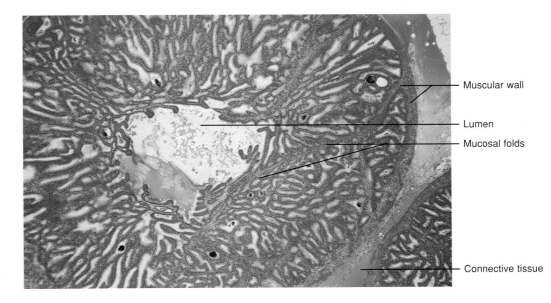

FIGURE 19.19

LM of a seminal vesicle in cross section, at a higher magnification. The micrograph
shows an intense irregular network of primary and secondary mucosal folds and the compartments between the folds. Vesicular secretion can be seen in the lumen. (40×)

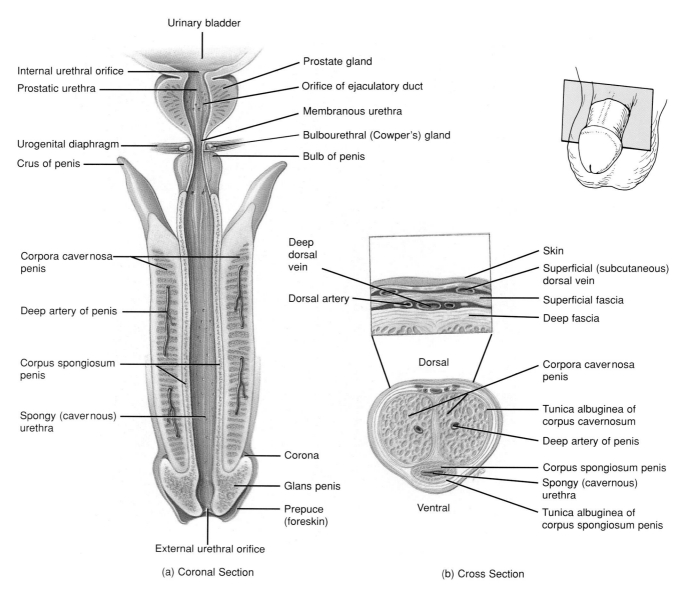

(a) Coronal Section

(b) Cross Section

FIGURE 19.20

(a) Diagram of a coronal section of the male reproductive organs, illustrating the general morphology. (b) Diagram of a cross section through the penis, illustrating the internal morphology.

FIGURE 19.21
LM of a cross section through a prostate gland lobule of a young male. The glandular stroma displays a dense cluster of fibroelastic connective tissue that also consists of numerous smooth muscle fibers. The epithelial lining in the gland may be simple columnar when the gland is inactive, or pseudostratified columnar to cuboidal when the gland is active. Also shown in the the micrograph are concentrations of prostate secretory products (prostatic concretions). (200×)

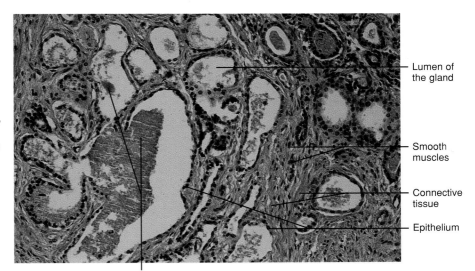

Lumen of the gland

Smooth muscles

Connective tissue

Epithelium

Prostatic concretions

FIGURE 19.22
LM of a cross section through the prostate gland (a tuboalveolar gland) of a young male, at a higher magnification. The micrograph shows small and large irregular cavities lined by cuboidal epithelium. Fibroelastic connective tissue surrounds the secretory alveoli. Vast amounts of smooth muscle fibers can be seen between the fibroelastic fibers. Also shown are prostatic concretions in the alveoli. (400×)

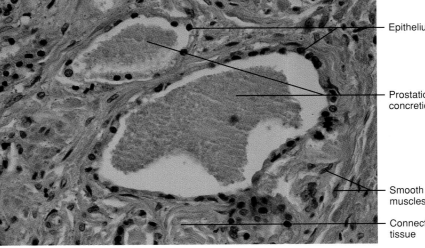

Epithelium

Prostatic concretions

Smooth muscles

Connective tissue

FIGURE 19.23
LM of a cross section through the prostate gland of an older man. As shown in the micrograph, the glandular epithelium is low columnar to low pseudostratified columnar. The fibroelastic fibers and smooth muscle bundles are sparse, and the lumen of one of the alveoli shows amorphous concentrations of corpora amylacea. (400×)

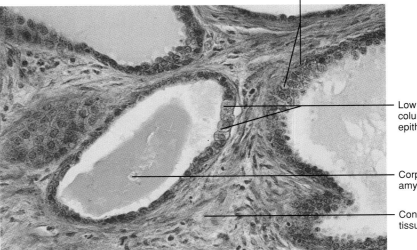

Low pseudostratified columnar epithelium

Low columnar epithelium

Corpora amylacea

Connective tissue

FIGURE 19.24
LM of a cross section through the penis of an infant primate at low magnification. Shown in the micrograph is the urethra surrounded by corpora cavernosa urethrae (initially called corpus spongiosum) and corpora cavernosa penis. Dense fibroelastic tissue surrounds the cavernous bodies. A Pacinian corpuscle (corpuscle of Vater-Pacini) can be seen in the dermis. Overlying the dermis is the epidermis of the prepuce or foreskin. (20×)

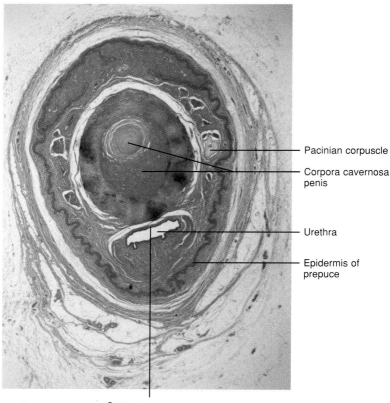

Pacinian corpuscle

Corpora cavernosa penis

Urethra

Epidermis of prepuce

Corpora cavernosa urethrae

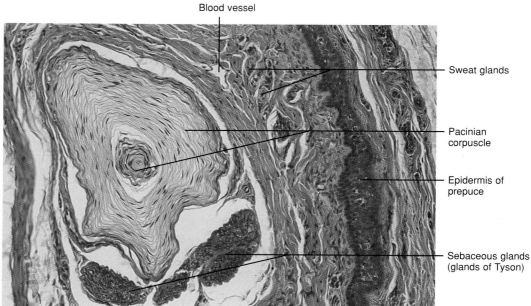

Blood vessel

Sweat glands

Pacinian corpuscle

Epidermis of prepuce

Sebaceous glands (glands of Tyson)

FIGURE 19.25
LM of a cross section through a small area of a primate penis, at a higher magnification. Shown in the micrograph is the skin covering the distal shaft of the penis (prepuce). Present in the underlying connective tissue are a corpuscle of Vater-Pacini and uncommon sebaceous glands (glands of Tyson) not associated with hair follicles. Also visible in the micrograph are sweat glands. (200×)

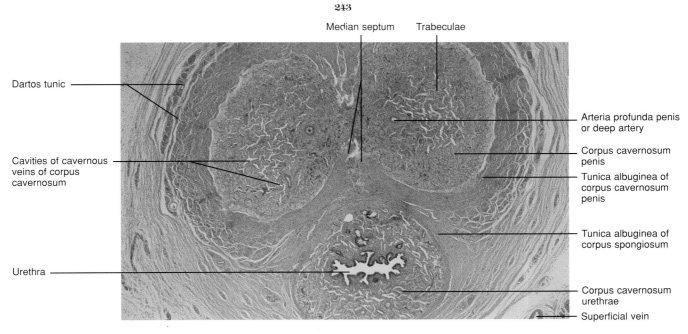

Median septum   Trabeculae

Dartos tunic

Cavities of cavernous
veins of corpus
cavernosum

Urethra

Arteria profunda penis
or deep artery

Corpus cavernosum
penis

Tunica albuginea of
corpus cavernosum
penis

Tunica albuginea of
corpus spongiosum

Corpus cavernosum
urethrae

Superficial vein

FIGURE 19.26
LM of a cross section through a human penis. The bulk of the penis consists of erectile tissue, the corpora cavernosa urethrae, and a pair of corpora cavernosa penis. The two masses of corpora cavernosa penis are united by a pectiniform septum. A thick fibrous sheath, the tunica albuginea, covers the erectile tissue. Also shown in the micrograph are a pair of deep arteries (arteria profunda penis) in the corpora cavernosa penis, the paraurethral glands, and the urethra in the corpus cavernosa urethrae. (20×)

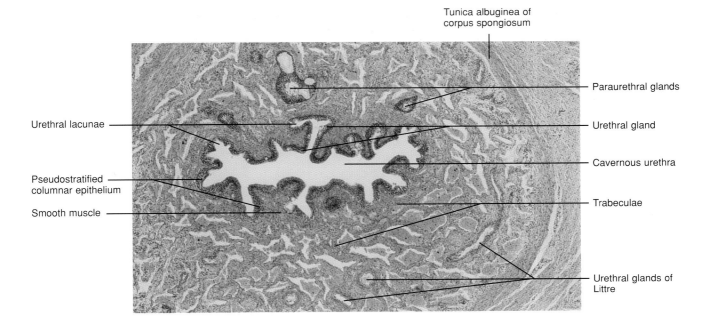

Tunica albuginea of
corpus spongiosum

Urethral lacunae

Pseudostratified
columnar epithelium

Smooth muscle

Paraurethral glands

Urethral gland

Cavernous urethra

Trabeculae

Urethral glands of
Littre

FIGURE 19.27
LM of a cross section through the corpora cavernosum urethrae (corpora spongiosum) of the penis. The micrograph shows the urethra and its pseudostratified columnar epithelial lining, the paraurethral glands, the smooth muscle, and the blood vessels associated with the surrounding connective tissue of the cavernosum urethrae. (40×)

FIGURE 19.28
LM of a cross section through the dorsal surface of the penis. The micrograph shows the dorsal artery, a large, deep dorsal vein, several small veins, the tunica albuginea, and the underlying corpus cavernosum penis. (100×)

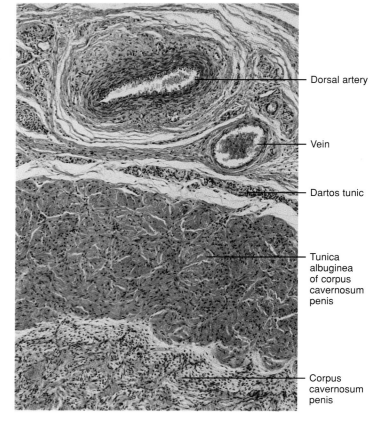

— Dorsal artery

— Vein

— Dartos tunic

— Tunica albuginea of corpus cavernosum penis

— Corpus cavernosum penis

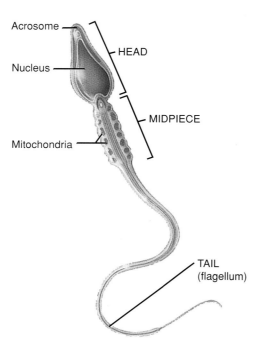

Acrosome

Nucleus

Mitochondria

HEAD

MIDPIECE

TAIL (flagellum)

FIGURE 19.29
Diagram showing the parts of a spermatozoa.

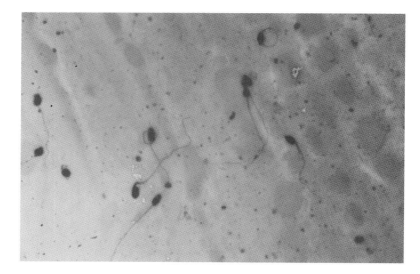

FIGURE 19.30
LM of human semen. The semen consists of spermatozoa mixed in seminal fluid that is a product of fluid from the testes, the seminal vesicles, the bulbourethral glands, and the prostate gland. (1000×)

The Female Reproductive System

The female reproductive system consists of the **ovaries, oviducts, uterus,** and **vagina.** The supporting external reproductive structures include the **labia majus, labia minus,** and **clitoris.** The **mammary glands** are also included in the female reproductive system, although they are not part of the genitalia.

Ovaries   The two ovaries are the oval bodies that lie on either side of the uterus, suspended by a mesentery from the broad ligament. Because of their secretions, the ovaries are classified as **endocrine** and **exocrine (cytogenic holocrine)** glands. As endocrine glands, the ovaries are the source of the cyclic secretion of the sex hormones, **estrogen** and **progesterone.** As exocrine glands, the ovaries produce a cyclic secretion of whole cells or **ova.**

In a cross section of the ovary, the **medulla** forms the inner portion, which is highly vascularized and contains little smooth muscle and few fibroelastic connective tissue fibers. The **cortex** of the ovary forms the outer layer, which consists of cellular stroma, an intense network of reticular fibers, and **ovarian follicles.** In the stroma of an active ovary, follicles in all stages of development are present. However, in a menopausal ovary, the cortical stroma becomes a narrow fibrous zone that lacks the ovarian follicles. In a given follicle, the immature ovum is surrounded by one or more layers of epithelial cells. As the follicle matures it undergoes several stages of development, progressing from a **primordial follicle** to a to **primary follicle,** then to a **secondary follicle,** and finally to a **mature (Graafian) follicle.**

In a developing follicle, a thin membrane, the **zona pellucida,** initially surrounds the ovum. In later stages of development, a mature follicle has an eccentrically placed ovum surrounded by **corona radiata** and **cumulus oophorus** cells that are displaced toward one side. The **antrum,** a large cavity filled with **liquor folliculi,** surrounds the ovum. Surrounding the antrum and the cumulus oophorus is a follicular capsule with an internal layer of cells, the **theca interna,** and an external layer of cells, the **theca externa.**

Fallopian Tubes or Oviducts   The two **fallopian tubes** connect the ovaries to the uterus. At the ovarian end, the oviduct is open and communicates with the peritoneal cavity. The other end of the oviduct opens into the **lumen** of the uterus. The oviduct can be divided into three segments. Close to the ovary, the tube flares into the **infundibulum,** a funnel-shaped structure with tapering **fimbriae.** Adjacent to the infundibulum is the **ampulla.** The ampulla forms half of the tube. The ampulla opens into the **isthmus** that terminates in the uterine lumen.

The mucosa of the ovary is highly folded, forming longitudinal folds known as the **plicae.** The epithelium is primarily simple columnar with an underlying lamina propria that is rich in reticulate fibers. The columnar cells are mixed ciliated and nonciliated. Thin inner circular and outer longitudinal layers of smooth muscle surround the mucosa.

Uterus   The uterus is a pear-shaped muscular organ, approximately 7 cm in length and 5 cm in width. The uterus can be divided into the **fundus** (rounded upper end of the body), the **corpus uteri** or **body** (the broad part of the uterus), and the **portio vaginalis,** a narrow cylindrical neck or cervix that projects into the vagina. The fallopian tubes enter the uterus at the **fundus.** The body wall of the uterus can be separated into three zones: the outer **serosa** or **perimetrium,** the middle **myometrium** or **muscularis,** and the inner **endometrium.** The perimetrium consists of a single layer of mesothelial cells, the myometrium is essentially thick layers of smooth muscles, and the endometrium, which undergoes cyclic changes, is lined by simple columnar epithelium with scattered ciliated cells. Underlying the epithelium is the connective tissue, the lamina propria, with uterine glands.

Cyclic Changes in the Endometrium   At puberty and thereafter, the endometrium goes through cyclic changes that end at menopause. The cyclic endometrial changes consist of a **menstrual** stage associated with a menstrual discharge, the **proliferation** (follicular) stage in which the follicle grows, matures, and releases the egg and the hormone estrogen, the **progestational** (luteal) stage associated with corpus luteum secretion, and finally the **ischemic** stage characterized by the cessation of blood flow to the coiled blood vessels of the endometrium.

Cervix   The cervix is the inferior segment of the uterus and is essentially a dense collagenous connective tissue. The body wall of the cervix lacks smooth muscle. The mucous membrane consists of tall mucus-secreting columnar cells mixed with some ciliated cells. The mucous lining forms deep furrows or clefts called **plicae palmatae.** The lamina propria is primarily cellular connective tissue with an absence of coiled blood vessels.

Vagina   The vagina forms the lowermost segment of the female reproductive tract. It is a fibromuscular sheath with a mucous membrane. Ordinarily, it is collapsed with anterior and posterior walls in contact. The body wall of the vagina is composed of the mucous layer, which forms transverse folds, or **rugae,** lined by thick, nonkeratinizing, stratified squamous epithelium. The muscularis contains smooth muscle fibers that are arranged in bundles. The adventitia forms a thin layer of connective tissue that blends with the surrounding tissue.

Mammary Glands   Both sexes have mammary glands; however, owing to the lack of female hormones, the glands in the male remain rudimentary throughout life. The mammary glands are modified sweat glands located within the subcutaneous tissue. Each mammary gland is divided into 15 to 20 lobes separated and surrounded by connective tissue composed primarily of adipose cells. Each lobe is an independent gland with its own duct that opens at the apex of the **nipple.** Each lobe is further divided into many **lobules** separated by connective and adipose tissue. **Intralobular ducts** from the lobules drain into the **interlobular ducts** of the lobes. The interlobular duct from each lobe joins and forms a single duct, the **lactiferous** duct, which traverses through the nipple and dilates into a **lactiferous sinus** just before it terminates in the summit of the nipple.

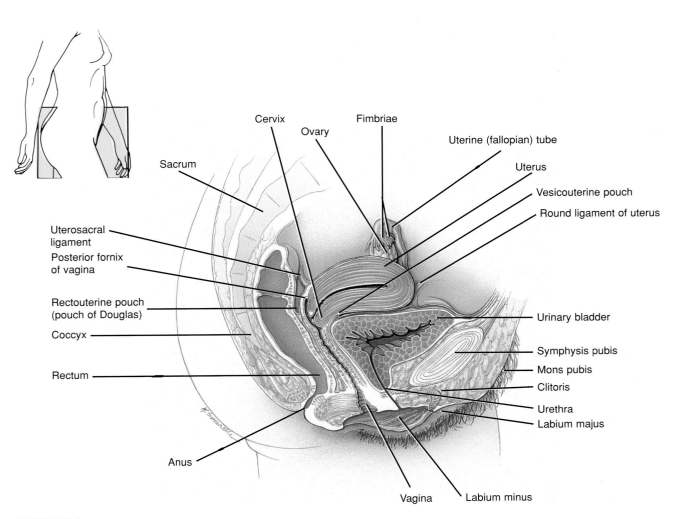

FIGURE 20.1
Diagram of the human female reproductive system and surrounding
organs in a sagittal section.

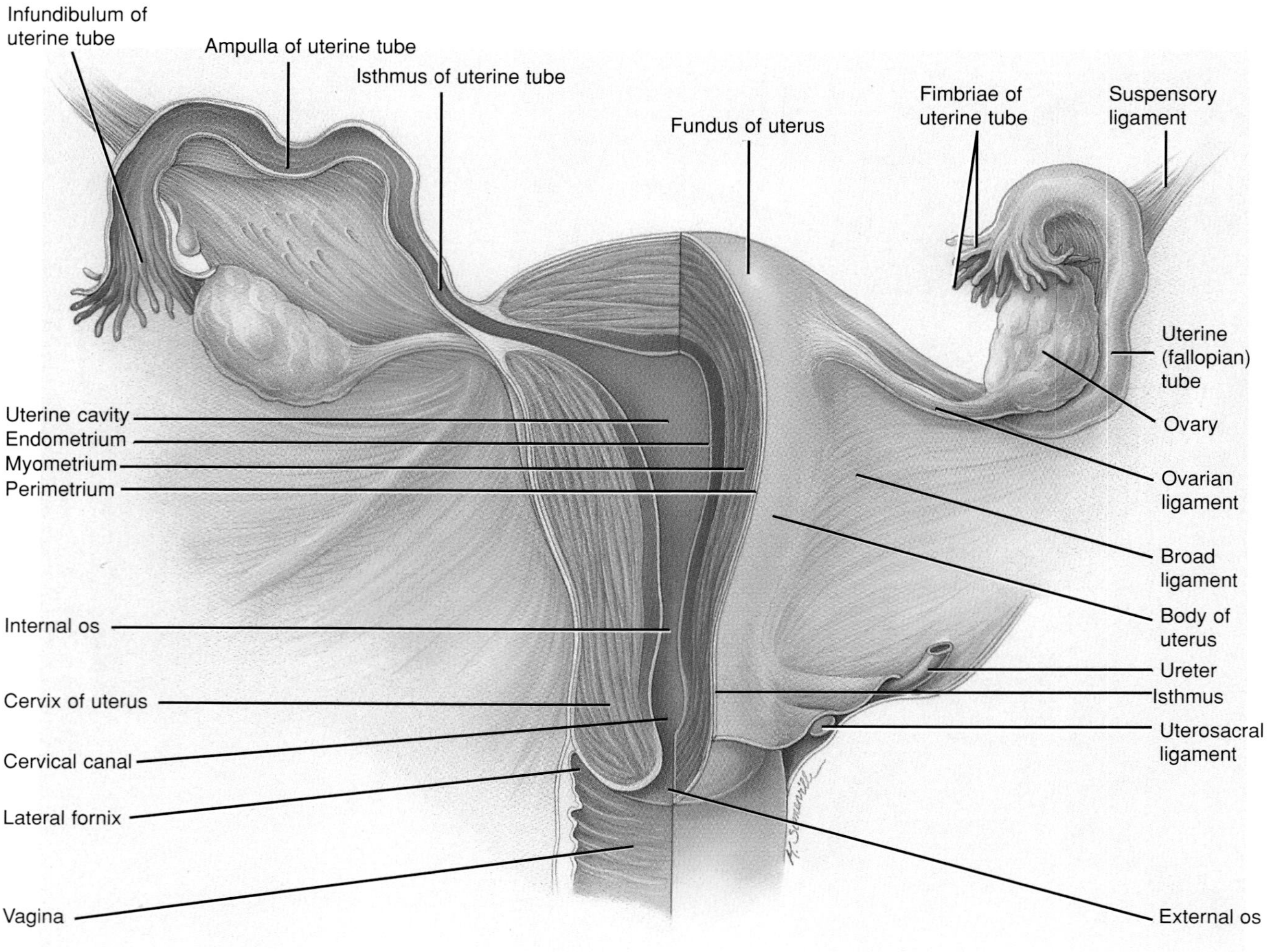

FIGURE 20.2
Diagram of the human female reproductive system as seen in a coronal section.

Ovary

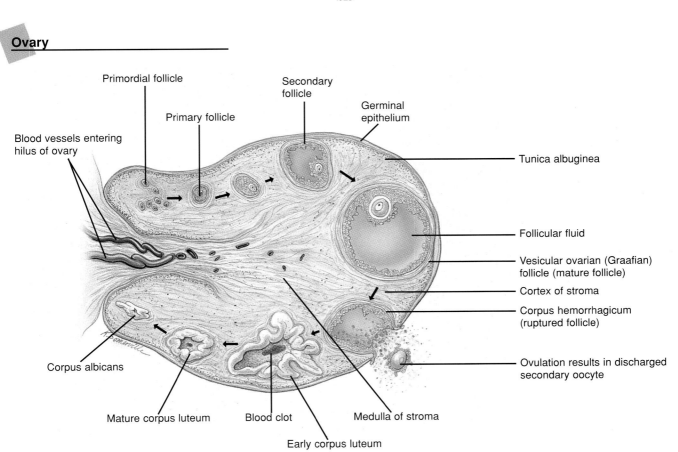

FIGURE 20.3
Diagram of the human ovary, illustrating the developmental stages associated with a mature oocyte.

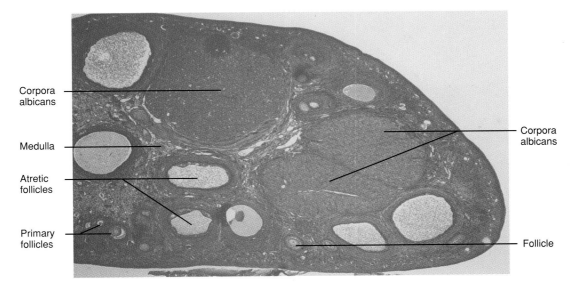

FIGURE 20.4
Light micrograph (LM) of a sagittal section through a cat ovary, illustrating postovulatory follicles of various types, follicles at different stages of development in the cortex area, and two distinct masses of corpora albicans. The medulla (central zone) of the ovary is highly vascular with coiled (helicine) arteries, veins, and lymph vessels. Also present in the ovary are nerves of the autonomic nervous system. (20×)

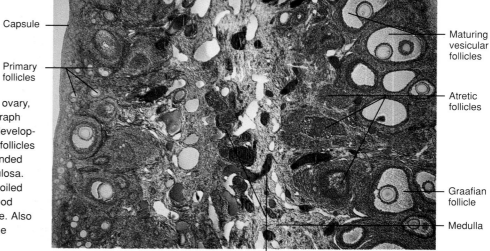

Capsule

Maturing vesicular follicles

Primary follicles

Atretic follicles

Graafian follicle

Medulla

Helicine arteries

FIGURE 20.5

LM of a sagittal section through a cat ovary, at a higher magnification. The micrograph shows follicles at different stages of development in the cortex area. Some of the follicles show a well-developed oocyte surrounded by a zona pellucida and a zona granulosa. The medulla (central zone) displays coiled arteries (helicine arteries), smaller blood vessels, and stromal connective tissue. Also present is the tunica albuginea and the cuboidal layer of the capsule. (40×)

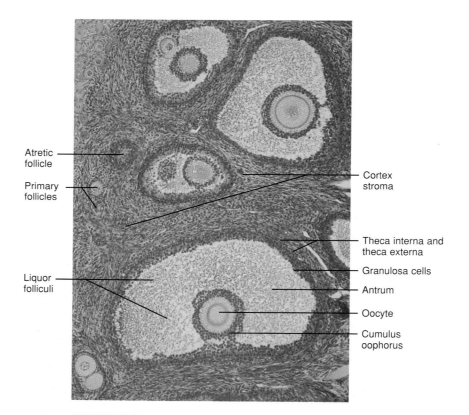

Atretic follicle

Primary follicles

Liquor folliculi

Cortex stroma

Theca interna and theca externa

Granulosa cells

Antrum

Oocyte

Cumulus oophorus

FIGURE 20.6

LM of a sagittal section through the cortex of an ovary, illustrating maturing follicles. The theca interna and theca externa, the follicular antrum, the liquor folliculi, the zona pellucida, the zona granulosa, and the surrounding stromal connective tissue of the ovary can be seen in the micrograph. (100×)

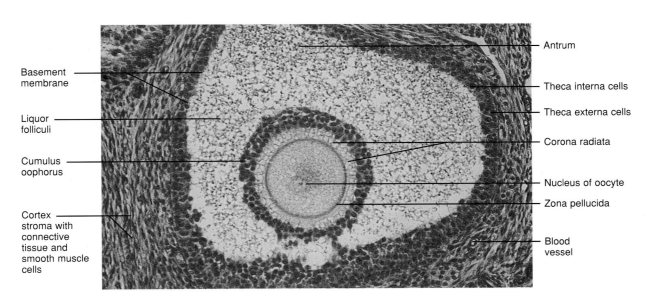

FIGURE 20.7

LM of a sagittal section through a Graafian follicle. The micrograph displays a mature oocyte surrounded by a zona pellucida, a corona radiata, and cells of the cumulus oophorus. A well-established follicular antrum with liquor folliculi, a theca interna, a theca externa surrounded by connective tissue fibers, and smooth muscle cells of the ovary stroma can be identified in the micrograph. (200×)

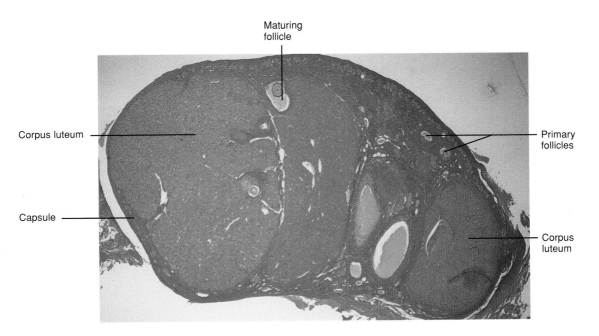

FIGURE 20.8

LM of a sagittal section through an ovary, demonstrating the development of a corpus luteum of pregnancy following ovulation, fertilization, and retention of the zygote in the endometrium. A few developing follicles are also visible in the micrograph. (20×)

Theca
lutein cells Cortical
stroma

Granulosa
lutein cells

Blood vessels

FIGURE 20.9
LM of a sagittal section through a developing corpus luteum. Shown in the micrograph are granulosa lutein cells. The cells are relatively large and contain numerous lipid droplets, which give the cells a vacuolated appearance. Surrounding the granulosa lutein cells is a thin layer of theca lutein cells. Also shown in the micrograph are blood vessels associated with the endocrine function of the corpus luteum. (100×)

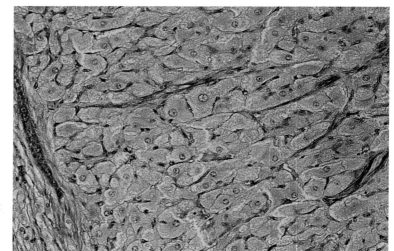

FIGURE 20.10
LM of a sagittal section through the corpus luteum cells, at a higher magnification. The cells have a finely vacuolated appearance resulting from lipid solvents used in the preparation of the specimen. (200×)

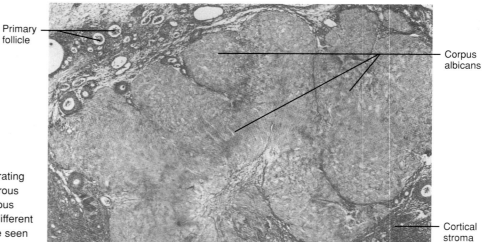

Primary
follicle

Corpus
albicans

Cortical
stroma

FIGURE 20.11
LM of an ovary in sagittal section, illustrating the corpus albicans and its acellular fibrous structure, after the involution of the corpus luteum. A few active follicular cells, at different stages of follicular development, can be seen surrounding the corpus albicans. (40×)

Fallopian Tubes (Oviducts)

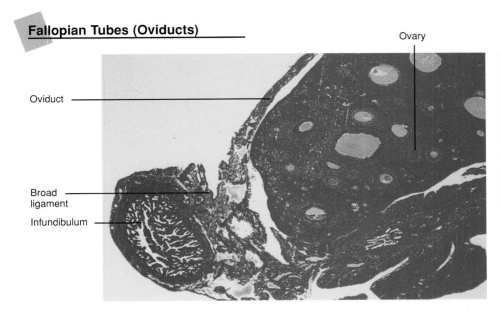

Ovary

Oviduct

Broad
ligament

Infundibulum

FIGURE 20.12
LM of a section through the ovary, the broad ligament, and the infundibulum of the oviduct. The mucosal lining of the infundibulum is highly convoluted and its margins form fringed folds, the fimbriae. The mucosal folds are supported by vascular connective tissue. The outer core of the oviduct consists of smooth muscle cells that are surrounded by dense connective tissue. (20×)

FIGURE 20.13
LM of a transverse section through the infundibulum of the oviduct (fallopian tube), at a higher magnification. Mucosal folds (fimbriae) extend deep into the lumen of the uterine tube. The epithelium lining is simple columnar with highly vascularized underlying lamina propria. The muscularis consists of an outer longitudinal muscle layer and an inner circular muscle layer. (100×)

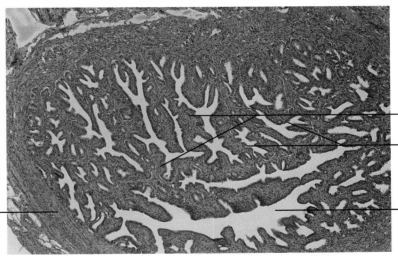

Fimbriae

Epithelial
lining

Lumen of
the oviduct

Muscularis

Lamina
propria

FIGURE 20.14
LM of a transverse section through the oviduct (fallopian tube) at the level of the ampulla. As the uterus is approached, the muscle wall of the oviduct thickens, the mucosal folds (fimbriae) decrease in number, and the epithelial lining progressively becomes ciliated columnar. (200×)

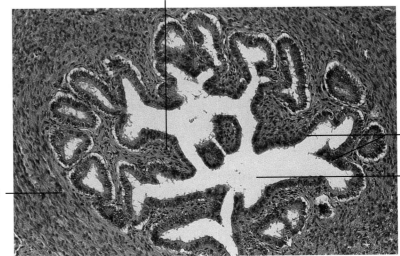

Columnar
cells

Lumen of the
oviduct

Muscularis

FIGURE 20.15

LM of a transverse section through the oviduct (fallopian tube) fimbriae, illustrating the mucosal lining. The epithelium consists of a single layer of columnar cells that may be ciliated or nonciliated. Most of the epithelial cells shown in the micrograph are ciliated cells. Below the epithelium lies the highly vascularized connective tissue layer, the lamina propria. (200×)

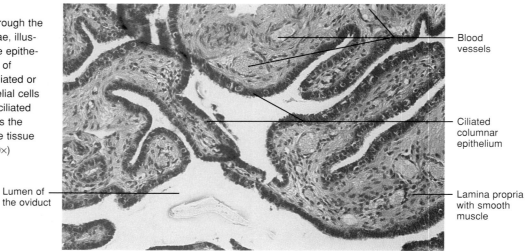

Blood vessels

Ciliated columnar epithelium

Lumen of the oviduct

Lamina propria with smooth muscle

Ciliated columnar cells

Lumen of the oviduct

Cilia

Layer of basal bodies

Lamina propria

FIGURE 20.16

LM of a cross section through a small portion of the oviduct (fallopian tube), at a higher magnification. Visible in the micrograph are the epithelial lining of ciliated columnar cells and the underlying connective tissue, the lamina propria. The cilia (kinocilia) in the oviduct have a distinct layer of basal bodies that presents itself in the form of a narrow bluish-black line. Such a line can be identified in the micrograph. (1000×)

Uterus

Stratum basalis

Uterine glands

Serosa

Myometrium

Lumen

Endometrium

Stratum spongiosum

Stratum compactum

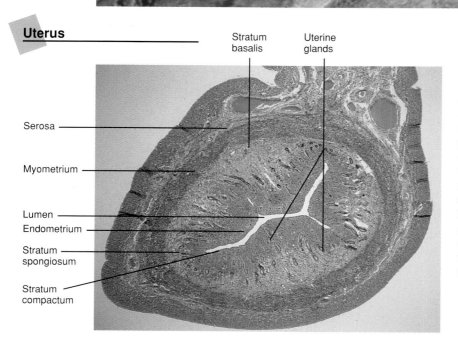

FIGURE 20.17

LM of a uterus in cross section, at low magnification. The micrograph shows the endometrium (mucous membrane), the smooth muscle layers of the myometrium, and an outer perimetrium or serosa. The endometrium is divided into three layers: the stratum basale, which is next to the myometrium; the stratum spongiosum, which is the middle layer; and the stratum compactum, which is the innermost layer next to the uterine lumen. Collectively, the stratum spongiosum and stratum compactum are called stratum functionalis because both of these layers are shed during menstruation. (40×)

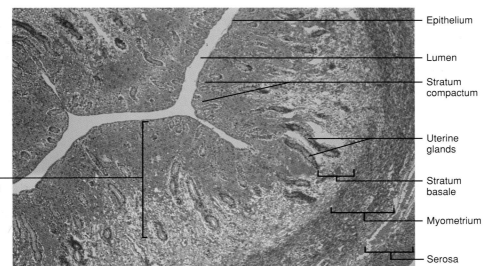

Epithelium

Lumen

Stratum
compactum

Uterine
glands

Stratum
basale

Myometrium

Serosa

Stratum
functionalis

FIGURE 20.18
LM of a uterus in cross section, at a higher magnification. The uterus is in the proliferative (follicular) phase. Shown in the micrograph is the columnar epithelium lining, the stratum compactum. The surface epithelium extends deep into the connective tissue of the lamina propria and forms the tubular uterine glands. Also visible in the micrograph are coiled spiral arteries in the endometrium, part of the muscular myometrium, and the outer perimetrium or serosa. (100×)

FIGURE 20.19
LM of a uterus in cross section, at a still higher magnification. The uterus is in the proliferative (follicular) stage. The lining consists of simple columnar epithelium that invaginates into the lamina propria and forms numerous long tubular glands called the uterine glands. Also shown in the micrograph are coiled spiral arteries associated with the stratum functionalis layer of the uterus. (200×)

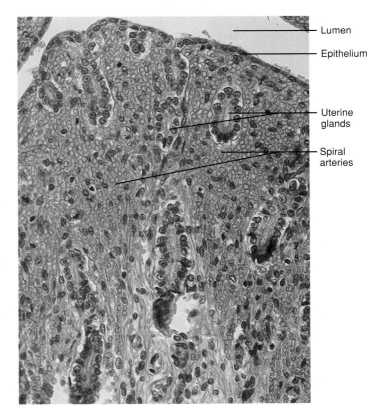

Lumen

Epithelium

Uterine
glands

Spiral
arteries

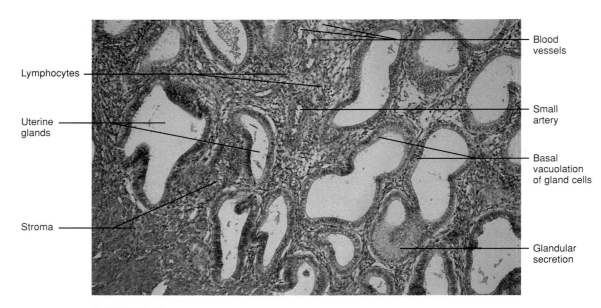

Lymphocytes

Uterine glands

Stroma

Blood vessels

Small artery

Basal vacuolation of gland cells

Glandular secretion

FIGURE 20.20

LM of a uterus in cross section, displaying the endometrium in the secretory or luteal phase of the menstrual cycle. There are increased glandular secretion and stromal edema. Lymphocytes can be seen in the stroma, which is highly vascularized. The uterine glands have hypertrophied and become tortuous as glycogen-rich secretory products accumulate in large quantities in their lumen. The basal cells display vacuolation, a characteristic feature of endometrium in the luteal stage. (100×)

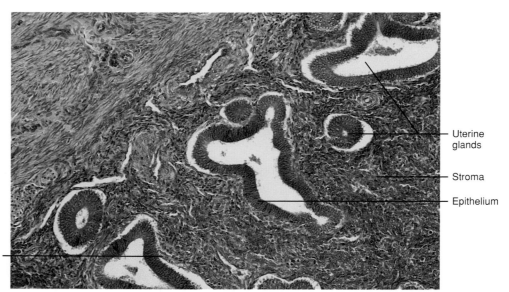

Uterine glands

Stroma

Epithelium

Lamina propria

FIGURE 20.21

LM of a cross section of endometrium in the late secretory or luteal stage of the menstrual cycle, at a higher magnification. The basal vacuolation of glandular cells is absent. The glands have become tortuous and separated from the lamina propria, and the stroma of the endometrium shows considerable edema. (200×)

Stroma

Erythrocytes

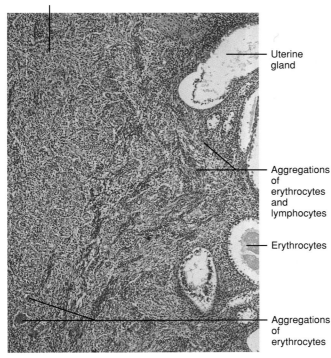

Uterine
gland

Aggregations
of
erythrocytes
and
lymphocytes

Erythrocytes

Aggregations
of
erythrocytes

Edema
in the
stroma

Disintegration
of epithelial
lining of the
uterine gland

Hemorrhaging
of
erythrocytes

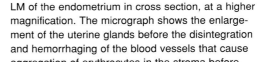

FIGURE 20.22
LM of the superficial layer of the endometrium in a
cross section at the onset of menstruation. The
blood vessels near the surface have ruptured,
adding blood to the glandular secretion. The
endometrial stroma of the functionalis displays
aggregations of erythrocytes and infiltration of
lymphocytes. (100×)

FIGURE 20.23
LM of the endometrium in cross section, at a higher
magnification. The micrograph shows the enlarge-
ment of the uterine glands before the disintegration
and hemorrhaging of the blood vessels that cause
aggregation of erythrocytes in the stroma before
menses. (200×)

Placenta

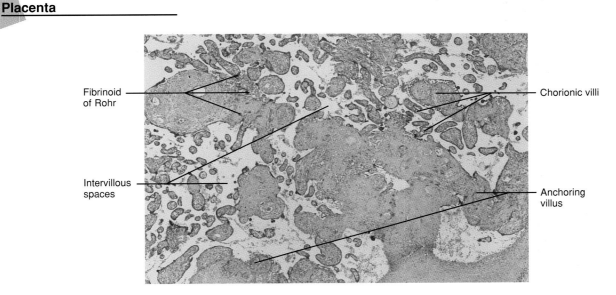

Fibrinoid
of Rohr

Chorionic villi

Intervillous
spaces

Anchoring
villus

FIGURE 20.24
LM of a cross section of a fetal portion of placenta, at low magnification. The
micrograph shows chorionic villi with a vascular lumen, intervillous spaces,
anchoring villus, and fibrinoid materials. (40×)

FIGURE 20.25
LM of a cross section of placental tissue, at a higher magnification. The micrograph shows several villi in cross section and the highly vascularized stroma of the villi. Also visible in the micrograph is the syncytiotrophoblast layer of a villus. Within the connective tissue of villi are intensely stained cells known as Hofbauer cells, which are closely associated with histiocytes. (100×)

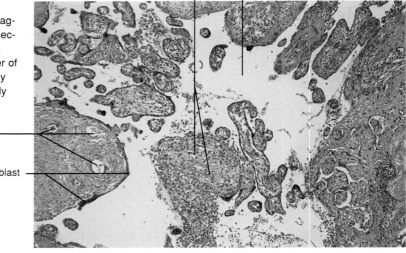

Hofbauer cells and leukocytes

Intervillous space

Fetal blood vessels

Syncytiotrophoblast

Cervix

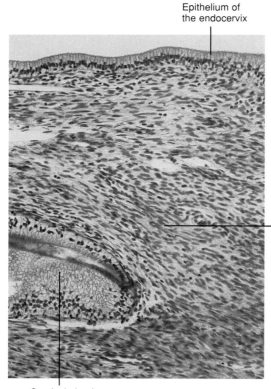

Epithelium of the endocervix

Lamina propria

Cervical gland

FIGURE 20.26
LM of a cross section through the inner body wall (endocervix) of the cervix at low magnification. Visible in the micrograph is the epithelium lined by mucus-secreting columnar cells, the cervical gland, and the connective tissue of the lamina propria. (200×)

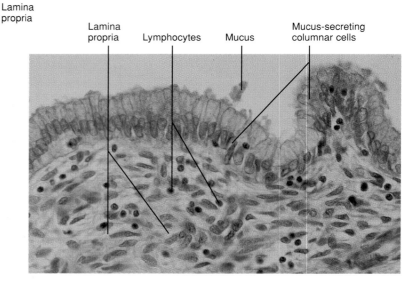

Lamina propria

Lymphocytes

Mucus

Mucus-secreting columnar cells

FIGURE 20.27
LM of a cross section through the endocervix epithelium, at a higher magnification. The micrograph shows tall, mucus-secreting columnar cells and the underlying lamina propria. The nuclei lie at the bases of the cells. Infiltrating the lamina propria are sparsely scattered lymphocytes. (400×)

Vagina and Cervix

Mucus

Mucus

Mucus-secreting gland

Vaginal-cervical junction

Vaginal epithelium

Lamina propria

Cervical fold

Cervical epithelium

Blood vessels

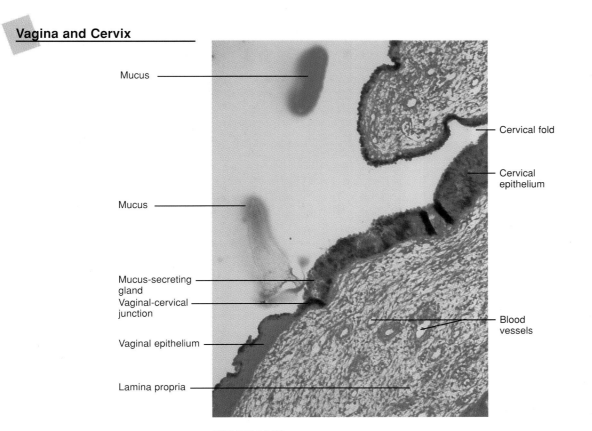

FIGURE 20.28

LM of a cross section through the vaginal-cervical junction (portio vaginalis). The cervical epithelium in the region is stratified squamous. In the micrograph, the cervical epithelium displays a mucus-secreting cervical gland in the active state. Continuous with the cervical epithelium is the stratified squamous epithelium of the vagina. Underlying the epithelia is the connective tissue of the lamina propria with numerous branching glands. (100×)

FIGURE 20.29

LM of a cross section through the external os (mouth) of the cervix, illustrating the stratified nonkeratinizing squamous epithelium. Visible in the micrograph is an active mucus-secreting gland. Underlying the epithelium is the fibrous lamina propria. (400×)

Lamina propria

Mucus

Gland

Cervical epithelium

Lymphocytes

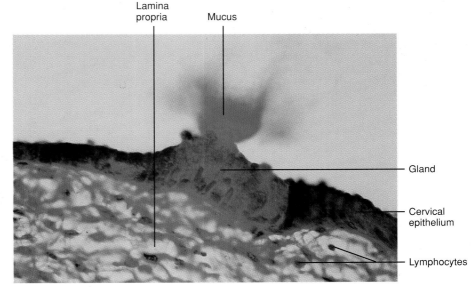

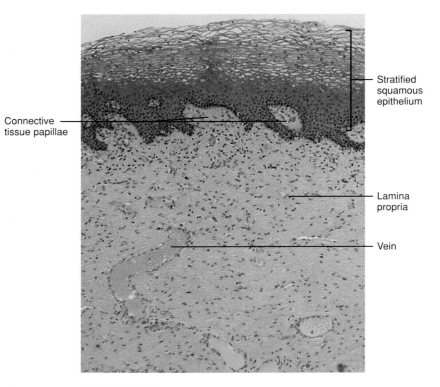

Stratified
squamous
epithelium

Connective
tissue papillae

Lamina
propria

Vein

FIGURE 20.30
LM of a cross section of the vagina, displaying the mucosal
layer lined by stratified squamous epithelium. Below the
epithelium lies the connective tissue of the lamina propria.
The lamina propria is devoid of glands but is rich in elastic
fibers. (100×)

FIGURE 20.31
LM of a cross section through the glycogen-rich,
stratified squamous epithelium of the vagina.
Desquamating epithelial cells can be seen in the
upper part of the epithelium. A narrow band of connective tissue, the lamina propria, underlies the
epithelium. (200×)

Lamina
propria

Desquamating
epithelial cells

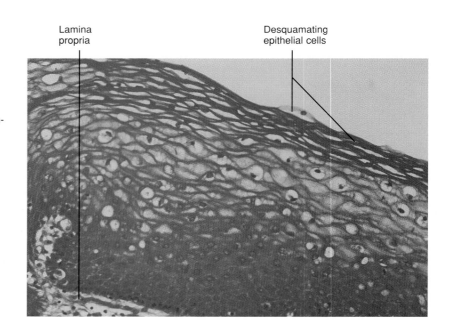

Mammary Glands

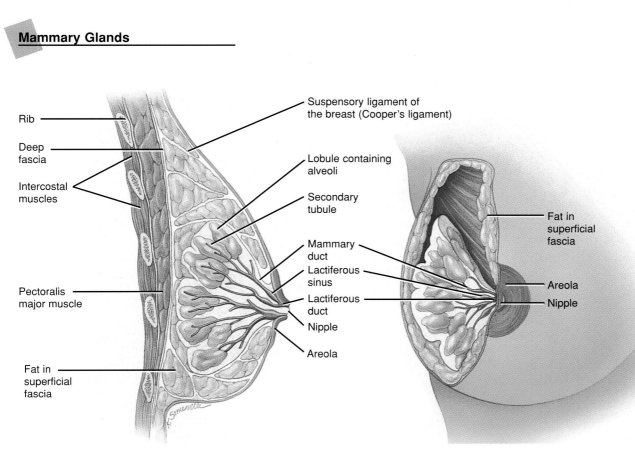

Rib

Deep fascia

Intercostal muscles

Pectoralis major muscle

Fat in superficial fascia

Suspensory ligament of the breast (Cooper's ligament)

Lobule containing alveoli

Secondary tubule

Mammary duct

Lactiferous sinus

Lactiferous duct

Nipple

Areola

Fat in superficial fascia

Areola

Nipple

FIGURE 20.32
Diagram illustrating the structure of the breast in frontal and sagittal planes.

FIGURE 20.33
LM of inactive mammary gland lobules in cross section, at low magnification. The micrograph shows dense interlobular collagenous connective tissue, blood vessels, islands of glandular tissue surrounded by dense fibrous and adipose tissue, and interlobular ducts. (100×)

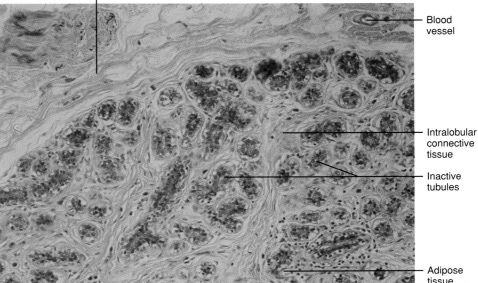

Interlobular connective tissue

Blood vessel

Intralobular connective tissue

Inactive tubules

Adipose tissue

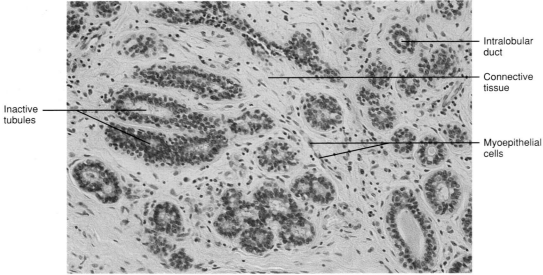

Inactive tubules

Intralobular duct

Connective tissue

Myoepithelial cells

FIGURE 20.34
LM of an inactive mammary gland lobule, at a higher magnification. Visible in the micrograph are myoepithelial cells surrounding the alveoli, small and large intralobular ducts lined by cuboidal cells, and intralobular loose connective tissue. (200×)

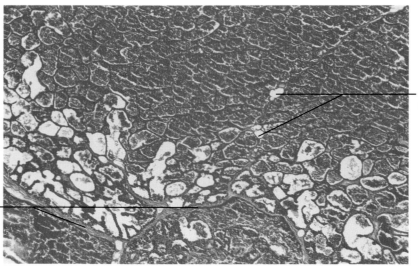

Intralobular ducts

Interlobular connective tissue

FIGURE 20.35
LM of a lactating breast in cross section. The micrograph shows active alveoli with secretion of milk. The alveoli are lightly packed, compressing the interalveolar fibrous connective tissue. Also shown in the micrograph are the interlobular connective tissue, inactive alveoli, and the intralobular ducts. (40×)

Organs of Special Sense

The organs of **sense** are distributed extensively in the epithelium, connective tissues, muscles, and tendons; however, special receptors associated with taste, sight, hearing, and smell are localized in the **tongue,** the **eyes,** the **ears,** and the **nose.** The taste buds of the tongue are described in Chapter 15 with the digestive system, and the nose is discussed in Chapter 16 with the respiratory system.

The Eye

The functioning of the eye has been compared to that of a camera; however, no such analogy can do justice to the relationship between the eye and the nervous system. The eye functions as a camera in that light from the external environment enters the eye through a lens, which focuses an image on the photosensitive cells of the **retina.**

The wall of the eyeball is a complex tissue structure. There are three identifiable layers: the outer **corneoscleral coat,** the middle **uveal layer,** and the inner **retina.** The uveal layer is the nutrient-rich middle vascular coat; it can be further divided into a posterior **choroid,** a **ciliary body,** and an anterior **iris.** The **retina**—the inner coat, which is considerably more complex than the other two coats—consists of photopigmented photoreceptors that convert light energy into chemical energy, which is then transformed into electrical energy by means of nerve impulses.

Internally, the eyeball has anterior and posterior chambers filled with a transparent watery nutrient fluid, the **aqueous humor,** and a vitreous chamber filled with **vitreous humor** (vitreous body), a viscous transparent gel.

The Corneoscleral Coat

The outer corneoscleral coat is composed of the **sclera,** an opaque, white fibrous coat in most of the posterior region, and the **cornea,** the transparent portion in the anterior region.

Sclera   The sclera forms a tough, fibrous outer tunic that covers five-sixths of the eyeball. The body wall of the sclera consists of flat bundles of collagen fibers mixed with elastic fibers. Between the fibers are flattened **fibroblasts. Melanocytes** are present in deeper layers. The sclera thins into a fenestrated membrane, the **lamina cribrosa,** at the point where the optic nerve exits the eyeball. The sclera functions as a site for the attachment of extrinsic muscles and as a protective coat for the eye.

Cornea   The cornea, the anteriormost portion of the eye, is a curved membrane (diameter of curvature, 11 mm) approximately 0.8 mm thick near the center and 1.0 mm thick at the periphery. The epithelium consists of four to five layers of large stratified squamous cells that form a smooth surface. The basal lamina of the epithelium rests upon an outer substantia propria, **Bowman's membrane,** that consists of fibers.

Descemet's Membrane   This uniform membrane lies between the lower part of the **substantia propria** and the corneal endothelium. This layer lacks elastic fibers but is resilient because of atypical collagen fibers.

The Corneal Endothelium   The corneal endothelium forms a single layer of cuboidal cells in the inner surface of the cornea, with the Descemet's layer serving as a basement membrane.

Limbus Corneae   The transitional zone between the cornea and the sclera is the limbus corneae. The shallow groove in the superior surface is the external scleral sulcus. Here the sclera blends with the cornea. The Bowman's and Descemet's membranes terminate in this area. A similar internal scleral sulcus lies on the inferior surface. Projecting from the internal scleral sulcus is the **scleral spur.** The **canal of Schlemm** lies between the **limbal stroma** and trabecular tissue, and is lined by flattened endothelial cells. The canal drains aqueous humor from the anterior chamber.

Middle Vascular Coat (Uveal Layer)

The vascular coat is divided into the **choroid,** the **ciliary body,** and the **iris.**

Choroid   The choroid is a highly vascular, brownish membrane adhering to the inner surface of the sclera by a thin, pigmented, elastic fiber layer called the suprachoroid lamina. Scattered in the elastic fibers are melanocytes and macrophages.

Ciliary Body   The ciliary body lies between the **ora serrata** of the **neural retina** and the outer surface of the iris. The bulk of the ciliary body is smooth muscle, which is adjacent to the sclera and an inner/vascular tunic that is lined by pigmented ciliary epithelium. The ciliary epithelium secretes aqueous humor. The **ciliary process** is the anterior inner surface of the ciliary body, and consists of 60 to 80 elongated ridges with melanocytes in their stroma.

Zonule Fibers   Zonule fibers are secreted by the nonpigmented ciliary epithelium. Their function is to hold the lens in place.

Iris   The iris is the anteriormost part of the vascular coat or **uvea.** The constrictor and dilator smooth muscle of the iris control the size of the pupillary opening. The iris, with its circular central aperture, the **pupil,** is suspended in the aqueous humor between the cornea and the lens. It divides the cavity in which it is suspended into an anterior chamber extending from the cornea to the iris and a narrow posterior chamber between the iris and the lens. The iris regulates the amount of light passing through the pupil. This enables the eye to maintain vision in a range of lighting conditions.

Retina

The innermost tunic of the eyeball is the retina. The anteriormost layer of the retina is lined by a non-sensitive iridial and ciliary layer. The posterior functional layers form the photoreceptor organs for the eye, extending from the anterior **ora serrata** to the posterior **optic papilla** of the optic nerve. A shallow depression, the **fovea centralis,** extends 2.5 mm from the optic papilla toward the temporal side. The **macula lutea,** or yellow spot, surrounds the fovea centralis.

In cross section, the retina has several layers from the external choroid to the internal optic nerve papilla. These layers are the pigmented epithelium, the layers of rods and cones, the external limiting membrane, the outer nuclear layer, the outer plexiform layer, the inner nuclear layer, the inner plexiform layer, the ganglion cell layer, the optic nerve fiber layer, and the internal limiting membrane.

The **accessory organs** for the eye include the **eyelids,** the **conjunctiva,** and the **lacrimal glands.**

Eyelids The eyelids are covered externally with skin and internally with a mucous membrane. The dermis contains sweat and sebaceous glands, skeletal muscles, elastic fibers, and small anterior hairs. The epithelium is nonkeratinized stratified squamous. The **glands of Moll** are the large modified sweat glands between and behind the eyelashes.

Conjunctiva The conjunctiva is a mucous membrane that lines the inner surface of the eyelids and reflects to cover the anterior surface of the eyeball and attach to the corneal epithelium at the corneal margin. The epithelial lining of the conjunctiva varies. At the edge of the cornea, the epithelium is stratified squamous with an underlying lamina propria.

Lacrimal Glands The lacrimal glands are located in the upper lateral corner of the eye orbit. These glands are serous and tubuloacinar, with myoepithelial cells surrounding the acini. From 10 to 15 excretory ducts drain into the conjunctival sac and the lacrimal passages. The lacrimal passages drain excess tears from the conjunctiva into the nasal cavity.

The Ear

The three divisions of the ear are the **external ear,** the **middle ear,** and the **inner ear.**

External Ear

The external ear includes the **auricle (pinna),** the **external auditory meatus,** the **ceruminous glands,** and the **cerumen.** The auricle is an irregular elastic cartilage covered with a **perichondrium.** The skin covers the perichondrium of the elastic cartilage. In humans, hair follicles, sweat glands, and sebaceous glands may be present in the dermis. The external auditory meatus is a long, narrow canal approximately 2.5 cm in length, with an irregular path from the auricle to the tympanic membrane. The skin lining of the meatus contains sebaceous and ceruminous glands that secrete cerumen, a bitter, protective secretion.

Middle Ear

The middle ear consists of the **tympanic cavity,** the **auditory ossicle,** the **tensor tympani,** the **stapedius,** the **auditory (eustachian) tube,** and the **tympanic membrane.** Also present in the middle ear is a series of canals and cavities in the endolymph-filled membranous labyrinths.

Tympanic Membrane The tympanic membrane is an obliquely oriented oval fibrous membrane at the end of the meatus, separating the external ear from the middle ear. The outer fibers are radial, and the inner fibers are circular.

Tympanic Cavity The tympanic cavity lies behind the tympanic membrane. The epithelial lining in the cavity is squamous or low cuboidal, shifting to columnar and ciliated at the opening of the **auditory** or **eustachian tube** in the middle ear. Present in the tympanic cavity are three bony ossicles: the **malleus,** the **incus,** and the **stapes.** Associated with the three ossicles are the **tensor tympani** and the **stapedius muscle.**

Auditory or Eustachian Tube The auditory tube is about 3.5 cm long and connects the **nasopharynx** with the tympanic cavity. The epithelium is ciliated columnar at the opening into the tympanic cavity and pseudostratified ciliated mixed with goblet cells near the nasopharynx. Present near the nasopharynx opening are **seromucous glands** in the underlying lamina propria.

Inner Ear

The inner ear includes the **bony labyrinth,** the **membranous labyrinth,** the **perilymph,** and the **endolymph.** The inner ear consists of canals and cavities within the fluid-filled **perilymph** in the **petrous region** of the temporal bone, the **osseous labyrinth.** The bony and membranous labyrinths have two functionally different components: the **vestibular labyrinth,** which controls sensory equilibrium, and a **cochlea** that consists of sensory neurons and special cells for hearing.

Vestibule The vestibule is medially located but separated from the tympanic cavity by a wall incorporating the **fenestra ovalis.** Posteriorly, the vestibule has three semicircular canals at right angles to each other: the anterior, posterior, and lateral canals. Each canal dilates at one end to form the **ampulla.** One end of the posterior canal fuses with the anterior canal and opens into the vestibule through the **crus commune.** The lateral canal also opens directly into the vestibule. Anteriorly, the lumen of the vestibule is continuous with the **cochlea,** a bony, spirally coiled tube.

Sensory nerve endings include the fibers from **cristae ampullares,** in the ampullae of the semicircular canals, the **maculae utriculi** in the utricle, and the **sacculi** in the saccule. The maculae utriculi and the sacculi are associated with static and kinetic senses. The hearing process is attributed to the organ of Corti, located on the **cochlear duct.**

Cochlea The cochlea is divided into three compartments: the upper **scala vestibuli,** the middle **scala media (cochlear duct),** and the lower **scala tympani.** The scala tympani and the scala vestibuli are **perilymphatic** spaces. The scala media or cochlear duct is associated with the endolymphatic system. The organ of Corti is located in this region and is supported by the vestibular membrane or **Reissner's membrane,** which is the basilar membrane of the cochlear duct. The organ of Corti consists of columnar supporting cells that are taller than the hair cells. The inner and outer **pillar cells,** the inner and outer **phalangeal cells,** the **border cells,** the **cells of Claudius,** and the **Henson's cells** are also part of the supporting cells of the organ of Corti. The organ of Corti transforms the vibrations that originate in the basilar membrane into nerve impulses.

The Eye

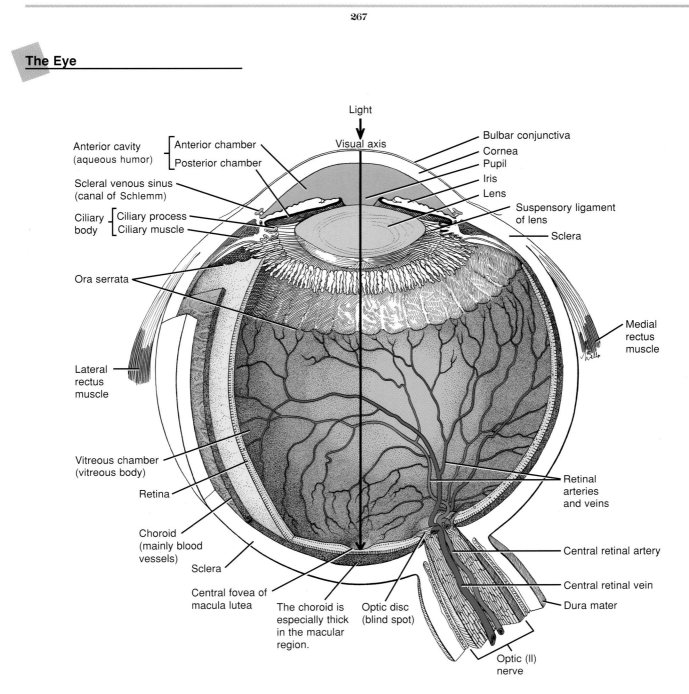

Light

Visual axis

Anterior cavity (aqueous humor) [Anterior chamber / Posterior chamber]

Scleral venous sinus (canal of Schlemm)

Ciliary body [Ciliary process / Ciliary muscle]

Ora serrata

Lateral rectus muscle

Vitreous chamber (vitreous body)

Retina

Choroid (mainly blood vessels)

Sclera

Central fovea of macula lutea

The choroid is especially thick in the macular region.

Optic disc (blind spot)

Bulbar conjunctiva

Cornea

Pupil

Iris

Lens

Suspensory ligament of lens

Sclera

Medial rectus muscle

Retinal arteries and veins

Central retinal artery

Central retinal vein

Dura mater

Optic (II) nerve

FIGURE 21.1
Diagram of a transverse section of the left eyeball (superior view),
illustrating the internal and external components of the eye.

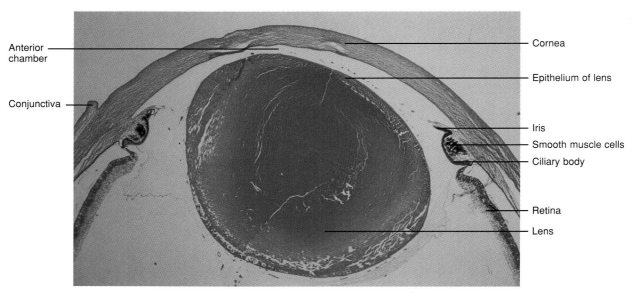

Anterior chamber

Conjunctiva

Cornea

Epithelium of lens

Iris

Smooth muscle cells

Ciliary body

Retina

Lens

FIGURE 21.2
Light micrograph (LM) of the anterior aspect of the human eye in a horizontal section. Illustrated in the micrograph are the lens, the ciliary bodies, the retina, the cornea, and part of the iris. (1×)

Elastic capsule

Epithelium

Fibers

FIGURE 21.3
LM of a transverse section through a small part of the crystalline lens of the eye. Three components of the lens can be identified in the micrograph: an outer elastic membranous capsule, a subcapsular epithelium of a single layer of cuboidal cells, and the lens substance composed of hexagonal lens fibers. The spaces between the fibers are artifacts. (400×)

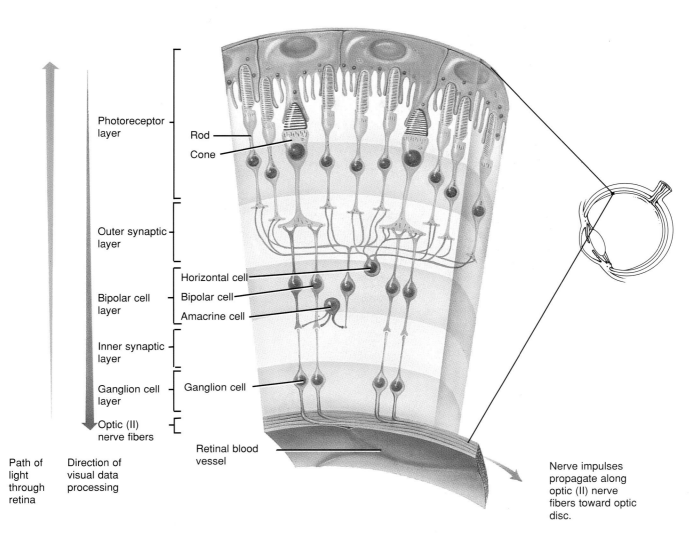

Photoreceptor layer

Rod

Cone

Outer synaptic layer

Bipolar cell layer

Horizontal cell
Bipolar cell
Amacrine cell

Inner synaptic layer

Ganglion cell layer

Ganglion cell

Optic (II) nerve fibers

Retinal blood vessel

Path of light through retina

Direction of visual data processing

Nerve impulses propagate along optic (II) nerve fibers toward optic disc.

FIGURE 21.4
Diagram of the microscopic structure of the retina, illustrating the downward signals passing through the neural portion of the retina (see arrow). Nerve impulses arise in the ganglion cells of the retina; from there they are transmitted to cranial nerve II, the optic nerve.

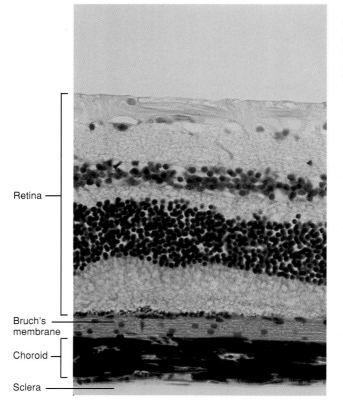

Retina —

Bruch's —
membrane

Choroid —

Sclera —

FIGURE 21.5

LM of a cross section through the body wall of the eye. The micrograph demonstrates the three prominent layers of the body wall. The layers are, from external to internal, the sclera, the choroid, and the retina. The choroid and retinal layers are separated by a thin membrane, the Bruch's membrane. (400×)

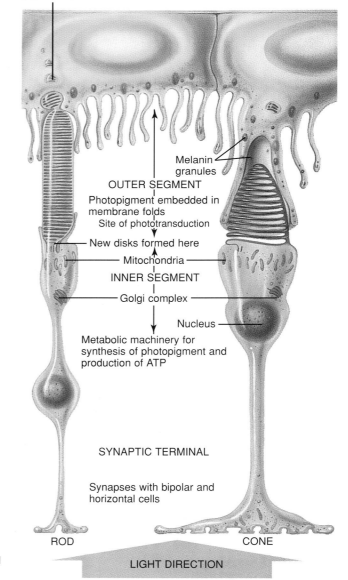

Old discs at tip slough off and are phagocytized by pigment epithelial cells.

Melanin granules

OUTER SEGMENT

Photopigment embedded in membrane folds
Site of phototransduction

New disks formed here

Mitochondria

INNER SEGMENT

Golgi complex

Nucleus

Metabolic machinery for synthesis of photopigment and production of ATP

SYNAPTIC TERMINAL

Synapses with bipolar and horizontal cells

ROD   CONE

LIGHT DIRECTION

FIGURE 21.6

Diagram illustrating the morphology of retinal photoreceptors, the rods and the cones.

Internal limiting membrane — 10
Optic nerve fiber layer — 9
Ganglion cell layer — 8

Inner plexiform layer — 7

Integrating bipolar cell layer — 6

Outer plexiform layer — 5

Cell bodies of rods and cones — 4

External limiting membrane — 3

Rods and cones — 2

Pigmented cells — 1

Choroid

Sclera

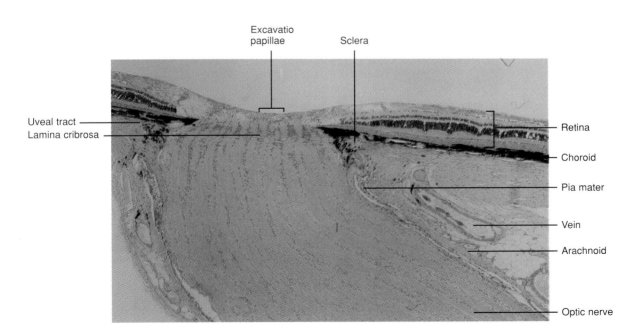

FIGURE 21.7
LM of a cross section through the body wall of the eyeball, demonstrating the two external layers, the sclera and the choroid, and subdivisions of the retina. In cross section, the retina can be divided into, from external to internal: (1) The pigmented epithelial cells; (2) the layer of rods and cones; (3) the external limiting membrane; (4) the cell bodies of rods and cones (outer nuclear layer); (5) the outer plexiform layer; (6) the integrating bipolar cell layer (inner nuclear layer); (7) the inner plexiform layer; (8) the ganglion cell layer; (9) the optic nerve fiber layer; and (10) the internal limiting membrane. (400×)

Excavatio papillae   Sclera

Uveal tract
Lamina cribrosa

Retina

Choroid

Pia mater

Vein

Arachnoid

Optic nerve

FIGURE 21.8
LM of the optic nerve and part of the retina and surrounding pia and arachnoid membranes of the meninges in a sagittal section. Shown in the micrograph are the retina, the excavatio papillae, the lamina cribrosa, the pia and the arachnoid membrane, and the optic nerve fibers. (40×)

FIGURE 21.9
LM of a cross section through the optic nerve, illustrating the central artery, the central vein, the pial connective tissue septa, and the nerve fiber bundles. (200×)

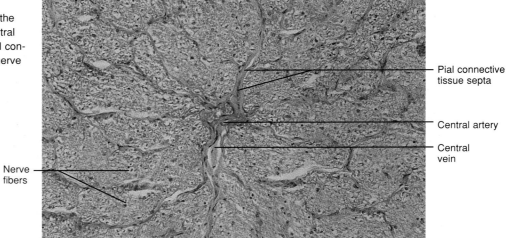

Pial connective tissue septa

Central artery

Central vein

Nerve fibers

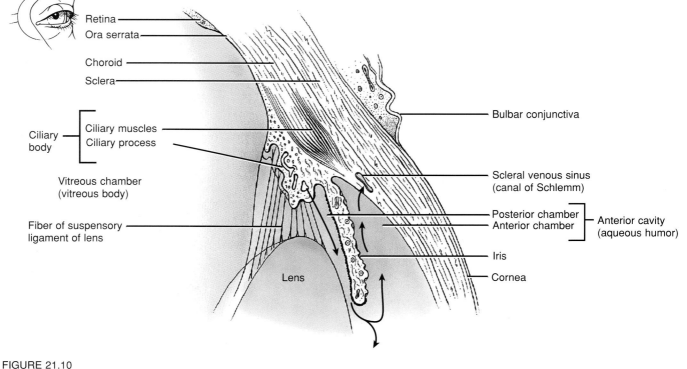

Retina

Ora serrata

Choroid

Sclera

Ciliary body
— Ciliary muscles
— Ciliary process

Vitreous chamber (vitreous body)

Fiber of suspensory ligament of lens

Lens

Bulbar conjunctiva

Scleral venous sinus (canal of Schlemm)

Posterior chamber
Anterior chamber
— Anterior cavity (aqueous humor)

Iris

Cornea

FIGURE 21.10
Diagram of a section through the anterior portion of the eyeball at the sclerocorneal junction. Aqueous humor flows in the direction indicated by the arrows.

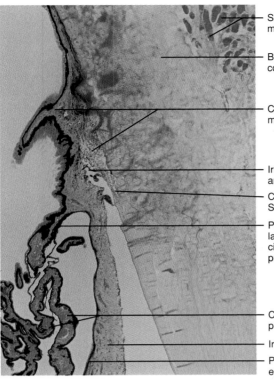

Skeletal muscle

Bulbar conjunctiva

Ciliary muscle

Iridocorneal angle

Canal of Schlemm

Pigmented layer of ciliary processes

Ciliary processes

Iris

Posterior iris epithelium

FIGURE 21.11
LM of a section through the ciliary body, the iris, and part of the bulbar conjunctiva. The micrograph illustrates the ciliary processes and their epithelial lining, the posterior iris and its epithelium, the ciliary muscle, the iridocorneal angle, and the canal of Schlemm. (100×)

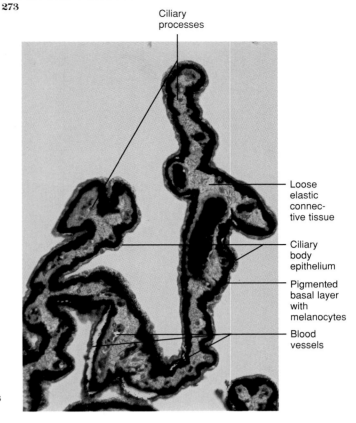

Ciliary processes

Loose elastic connective tissue

Ciliary body epithelium

Pigmented basal layer with melanocytes

Blood vessels

FIGURE 21.12
LM of a section through the pigmented ciliary body, at a higher magnification. The epithelium is a double layer of cuboidal cells supported by a basal lamina. Underlying the basal lamina are connective tissue and ciliary smooth muscle cells. Blood vessels can also be seen. (200×)

FIGURE 21.13
LM of an iris in a longitudinal plane. The iris is the anteriormost part of the uveal layer of the eye. The internal mass of the iris consists of highly vascular connective tissue with pigmented melanocytes. The anterior layer of the iris is irregular and displays a large number of fibroblasts mixed with melanocytes, giving a dark pigmented appearance. The posterior surface of the iris is smooth, nonpigmented, and lined by an epithelium that is a continuation of the ciliary body. (40×)

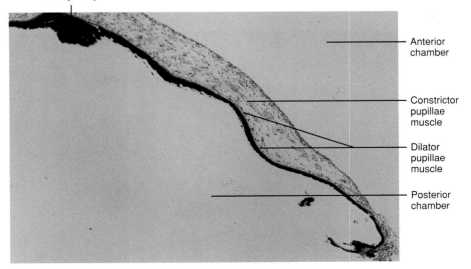

Ciliary Body

Anterior chamber

Constrictor pupillae muscle

Dilator pupillae muscle

Posterior chamber

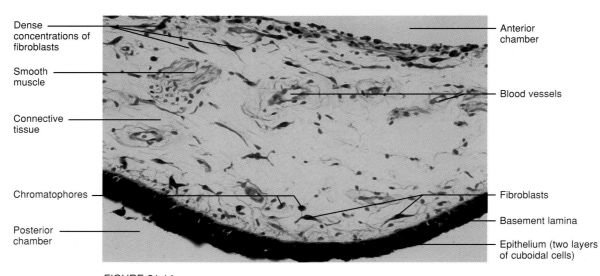

Dense concentrations of fibroblasts

Smooth muscle

Connective tissue

Chromatophores

Posterior chamber

Anterior chamber

Blood vessels

Fibroblasts

Basement lamina

Epithelium (two layers of cuboidal cells)

FIGURE 21.14

LM of a section through a small portion of an iris, at a higher magnification. The posterior surface is highly pigmented and consists of shallow furrows. Collagen fibers, fibroblasts, scattered pigmented cells, and smooth muscle form the highly vascular stroma of the iris. (200×)

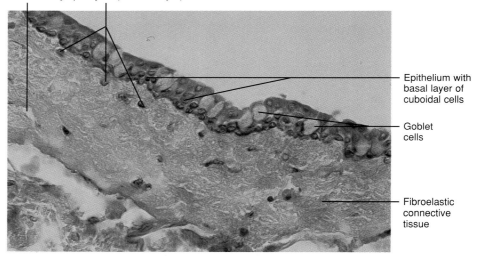

Vein   Lymphocytes (adenoid layer)

Epithelium with basal layer of cuboidal cells

Goblet cells

Fibroelastic connective tissue

FIGURE 21.15

LM of a section through a portion of the conjunctiva. This is the area of the conjunctiva where the stratified (bilayer) low columnar or cuboidal cells form an epithelial lining in which the surface cells are conic or cylindrical in shape. Interspersed among the epithelial cells are mucus-secreting cells. Underlying the epithelium is the lamina propria of fine fibroelastic connective tissue. (400×)

Conjunctiva

Corneal epithelium

Cornea

FIGURE 21.16

LM of a section through the conjunctiva and the peripheral surface of the cornea (limbus) at the transitional or junctional zone. At this junctional zone, the corneal epithelium thickens and becomes continuous with the conjunctiva. (200×)

FIGURE 21.17
LM of a cross section through the cornea of the eye. The cornea is avascular and transparent. As illustrated in the micrograph, the cornea consists of five distinct layers: (1) the epithelium; (2) the Bowman's membrane; (3) the substantia propria; (4) the Descemet's membrane; and (5) the cuboidal endothelium. Descemet's membrane is the basement membrane for the endothelium. (200×)

Anterior epithelium

Bowman's membrane

Substantia propria

Posterior epithelium (endothelium)

Descemet's membrane

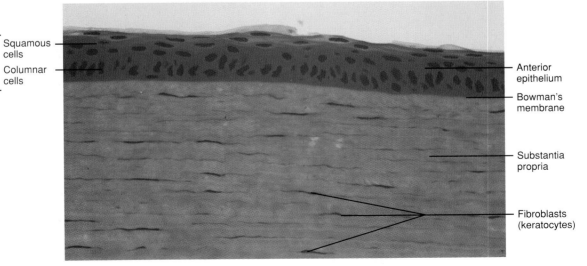

Stratified squamous epithelium

Squamous cells

Columnar cells

Anterior epithelium

Bowman's membrane

Substantia propria

Fibroblasts (keratocytes)

FIGURE 21.18
LM of a cross section through the outer three layers of the cornea, at a higher magnification. The micrograph shows the external stratified squamous nonkeratinizing epithelium. Underlying the epithelium is the acellular Bowman's membrane. Below the Bowman's membrane lies the substantia propria, which forms the bulk of the cornea. (400×)

FIGURE 21.19
LM of a cross section through the eyelid. The micrograph illustrates the supporting plate of connective tissue and a thin layer of skeletal muscle (orbicularis oculi) lined internally by a mucous membrane. Externally, the muscle is covered by stratified, nonkeratinizing squamous epithelium. Anteriorly, the eyelid is thin with small hairs, sebaceous glands, and sweat glands. The dermis at the margin of the eyelid is thick and displays three or four rows of long, stiff hairs, the eyelashes. (40×)

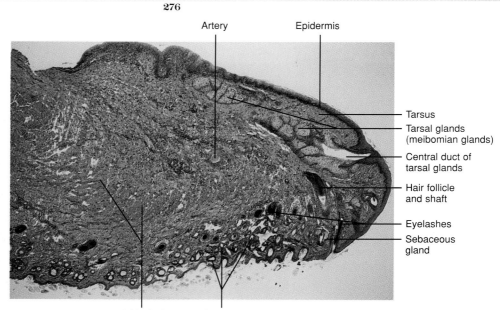

Artery   Epidermis

Tarsus

Tarsal glands (meibomian glands)

Central duct of tarsal glands

Hair follicle and shaft

Eyelashes

Sebaceous gland

Orbicularis oculi muscle   Sweat glands (of Moll)

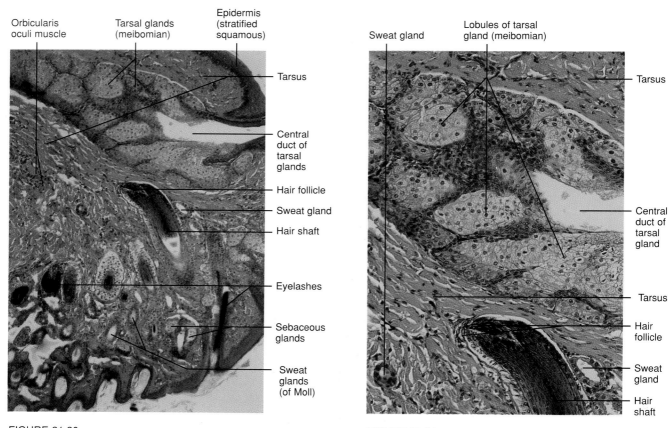

Orbicularis oculi muscle   Tarsal glands (meibomian)   Epidermis (stratified squamous)

Tarsus

Central duct of tarsal glands

Hair follicle

Sweat gland

Hair shaft

Eyelashes

Sebaceous glands

Sweat glands (of Moll)

Sweat gland   Lobules of tarsal gland (meibomian)

Tarsus

Central duct of tarsal gland

Tarsus

Hair follicle

Sweat gland

Hair shaft

FIGURE 21.20
LM of a section through the eyelid, at a higher magnification. The micrograph illustrates the uneven, thick dermal margin of the eyelid, several hair follicles (some with hair shafts), sweat glands (of Moll), sebaceous glands associated with eyelashes, striated muscle fibers of orbicularis oculi, and large tarsal glands (meibomian). The ducts from the tarsal glands open at the eyelid margin. (100×)

FIGURE 21.21
LM of a section through the eyelid, illustrating modified sebaceous tarsal (meibomian) glands in the tarsal plate. A hair shaft, skeletal muscle, connective tissue, an excretory duct, and a blood vessel can also be identified in the micrograph. (200×)

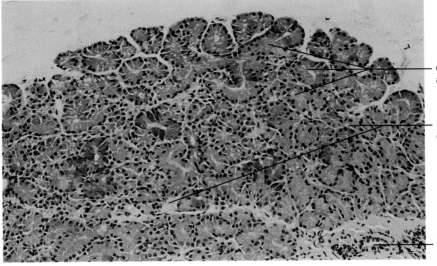

FIGURE 21.22
LM of a cross section through the lacrimal gland and ducts (lacrimal apparatus) that drain into the conjunctival sac. The lacrimal gland is a tear-secreting tubuloacinar and serous gland in which the lobules are surrounded by myoepithelial cells. Visible in the micrograph are lobules with secretory units. The nuclei in the cells are basally located. (200×)

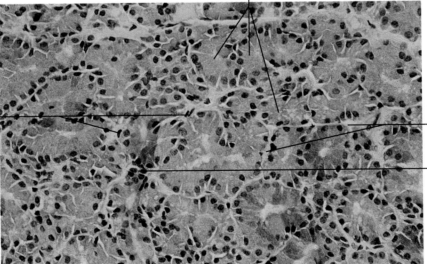

FIGURE 21.23
LM of a cross section through the lacrimal gland, at a higher magnification. The micrograph demonstrates the compound tubuloacinar secretory units of the gland. The acini are separated by a thin connective tissue septa. Tears (aqueous secretion), after bathing the conjunctiva-corneal surface, drain into collecting ducts, from which they are transported to the inferior meatus of the nose via the nasolacrimal duct. (400×)

The Ear

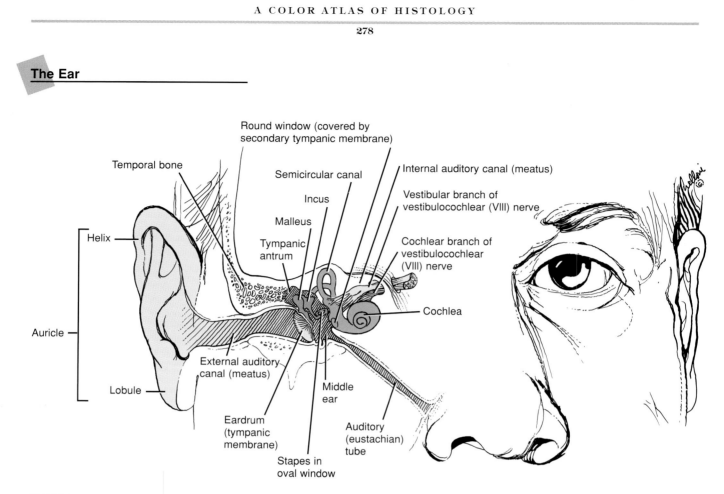

FIGURE 21.24

Diagram of the morphology of the ear as seen in a frontal section through the right side of the skull. The diagram shows the components of the external ear, the middle ear, and the inner ear.

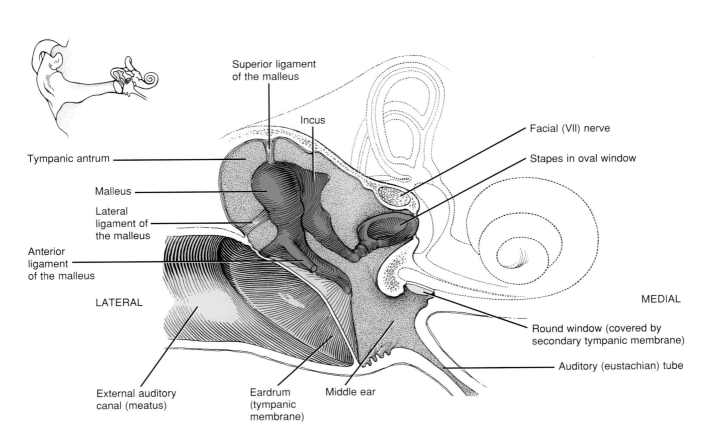

Superior ligament
of the malleus

Incus

Facial (VII) nerve

Stapes in oval window

Tympanic antrum

Malleus

Lateral
ligament of
the malleus

Anterior
ligament
of the malleus

LATERAL

MEDIAL

Round window (covered by
secondary tympanic membrane)

Auditory (eustachian) tube

External auditory
canal (meatus)

Eardrum
(tympanic
membrane)

Middle ear

FIGURE 21.25
Diagram of the auditory ossicles (malleus, incus, and stapes) located
in the middle ear and their relationship to the tympanic membrane
and the external auditory canal (meatus). Also shown in the diagram
are components of the inner ear that consist of the cochlea and its
internal structures.

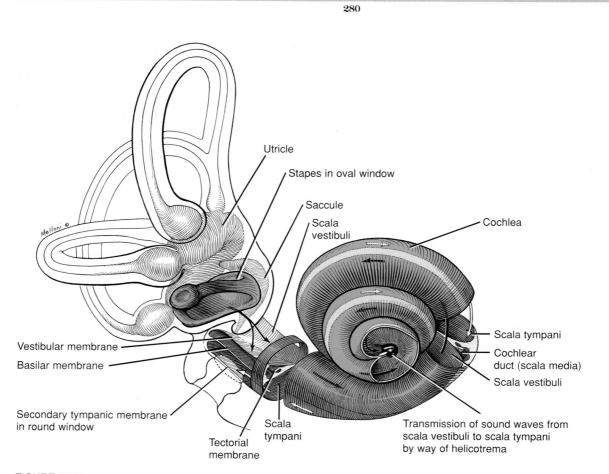

Utricle

Stapes in oval window

Saccule

Scala vestibuli

Cochlea

Vestibular membrane

Basilar membrane

Scala tympani

Cochlear duct (scala media)

Scala vestibuli

Secondary tympanic membrane in round window

Scala tympani

Tectorial membrane

Transmission of sound waves from scala vestibuli to scala tympani by way of helicotrema

FIGURE 21.26
Diagrammatic representation of the semicircular canals, the utricle, the saccule, the vestibule, and the cochlea. Note that the cochlea makes approximately three turns.

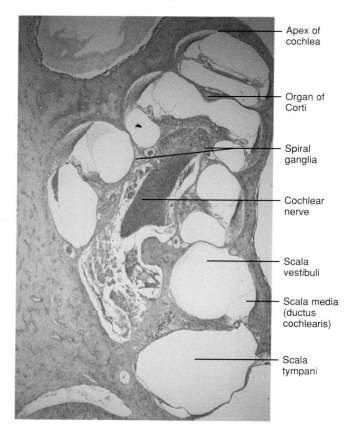

Apex of cochlea

Organ of Corti

Spiral ganglia

Cochlear nerve

Scala vestibuli

Scala media (ductus cochlearis)

Scala tympani

FIGURE 21.27
LM of the vestibulocochlear apparatus in a horizontal section. Shown in the micrograph are the bony cochlea, the helicotrema, the cochlear duct, the vestibulocochlear nerve, the spiral ganglia, the spiral organ of Corti, the scala tympani, the scala vestibuli, and the vestibule. A small portion of the middle ear can be seen at the top left. (20×)

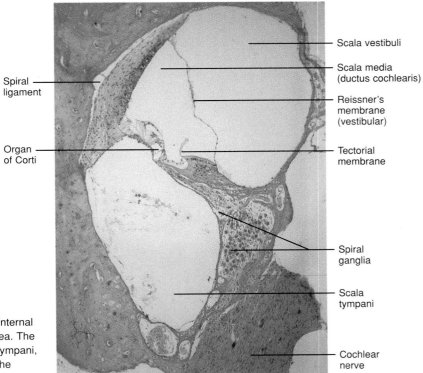

Spiral ligament

Organ of Corti

Scala vestibuli

Scala media (ductus cochlearis)

Reissner's membrane (vestibular)

Tectorial membrane

Spiral ganglia

Scala tympani

Cochlear nerve

FIGURE 21.28
LM of the cochlea in section, demonstrating the internal structures as seen in one turn of the spiral cochlea. The micrograph shows the scala vestibuli, the scala tympani, the spiral organ of Corti, the spiral ganglia, and the vestibular membrane. (200×)

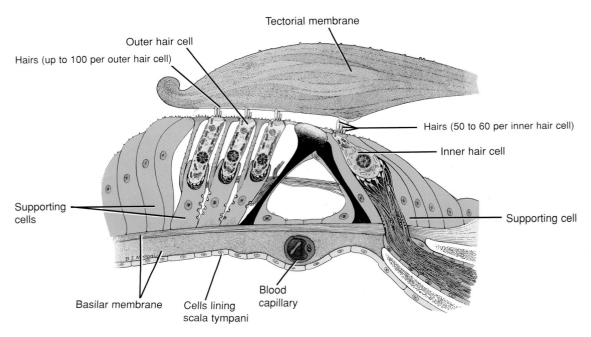

Tectorial membrane

Outer hair cell

Hairs (up to 100 per outer hair cell)

Hairs (50 to 60 per inner hair cell)

Inner hair cell

Supporting cells

Supporting cell

Basilar membrane

Cells lining scala tympani

Blood capillary

FIGURE 21.29
Diagram of the organ of Corti (spiral organ), illustrating its morphological components. The organ lies between the scala tympani and the scala vestibuli of the bony cochlea.

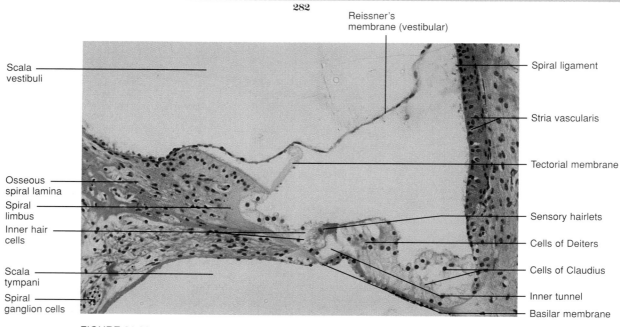

Reissner's
membrane (vestibular)

Scala
vestibuli

Spiral ligament

Stria vascularis

Tectorial membrane

Osseous
spiral lamina

Spiral
limbus

Inner hair
cells

Sensory hairlets

Cells of Deiters

Cells of Claudius

Scala
tympani

Spiral
ganglion cells

Inner tunnel

Basilar membrane

FIGURE 21.30

LM of a section through the organ of Corti in the cochlear duct, and parts of the scala tympani and scala vestibuli. Shown in the micrograph are the spiral ganglia, the osseous spiral lamina, the basilar membrane, the spiral ligament, the vestibular ligament, the tectorial membrane, the cochlear nerve fibers, and the epithelial lining of the cochlear duct. The organ of Corti is located in the cochlear duct. (200×)

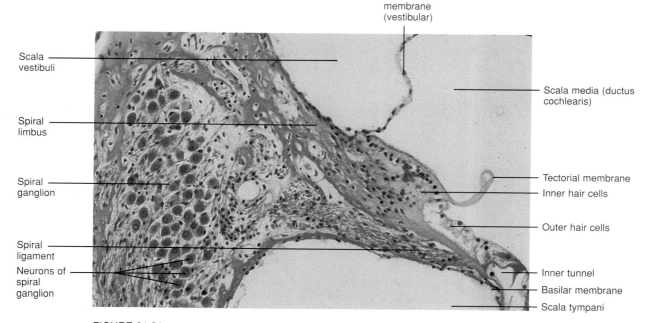

Reissner's
membrane
(vestibular)

Scala
vestibuli

Scala media (ductus
cochlearis)

Spiral
limbus

Spiral
ganglion

Tectorial membrane

Inner hair cells

Outer hair cells

Spiral
ligament

Neurons of
spiral
ganglion

Inner tunnel

Basilar membrane

Scala tympani

FIGURE 21.31

LM of a section through the spiral organ of Corti and the spiral ganglion. The micrograph shows the vestibular membrane, the tectorial membrane, the basilar membrane, the cochlear nerve fibers, part of the scala vestibuli, the scala tympani, and the cochlear duct. Also identifiable are supporting cells, outer hair (outer phalangeal) cells, inner phalangeal cells, pillar cells, an artery, veins, and the tunnel of the organ of Corti. (200×)

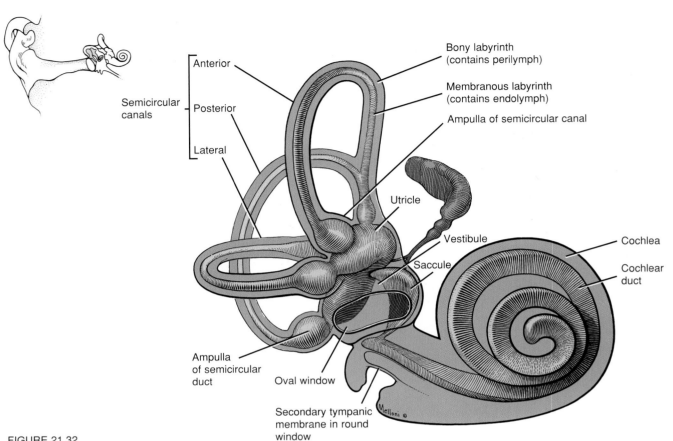

FIGURE 21.32
Diagram of the inner ear, illustrating the
bony labyrinth with the internal membranous
labyrinth, the utricle, the saccule of the
vestibule, the cochlea with the internal
cochlear duct, and the endolymphatic sac.

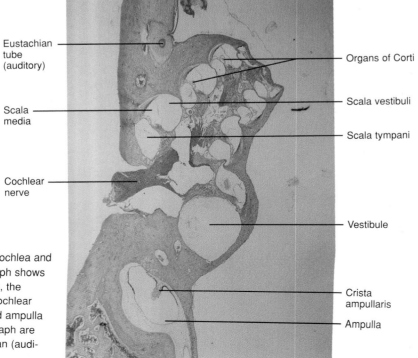

FIGURE 21.33
LM, at low magnification, of a section through the cochlea and
the ampulla of the semicircular canal. The micrograph shows
the bony or osseous labyrinth of the cochlea spirals, the
cochlear nerve, several organs of Corti within the cochlear
duct, and the crista ampullaris that lies in the dilated ampulla
of the semicircular canal. Also visible in the micrograph are
the cochlear nerve and the opening of the eustachian (auditory) tube. (20×)

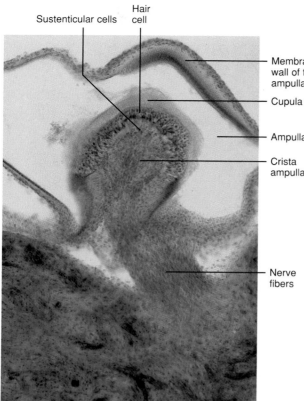

Sustenticular cells — Hair cell

Membranous wall of the ampulla

Cupula

Ampulla

Crista ampullaris

Nerve fibers

FIGURE 21.34

LM of a sagittal section through the dilated ampullary section of the semicircular canal. Within the ampulla lies the crista ampullaris, which functions as the receptor organ of the middle ear. A gelatinous cupula (capsule) surrounds the free end of the crista. Nerve fibers innervate the connective tissue stroma of the crista ampullaris. (100×)

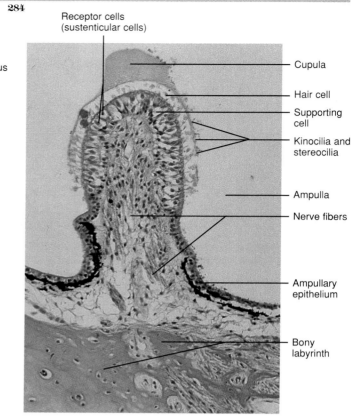

Receptor cells (sustenticular cells)

Cupula

Hair cell

Supporting cell

Kinocilia and stereocilia

Ampulla

Nerve fibers

Ampullary epithelium

Bony labyrinth

FIGURE 21.35

LM of a section through the crista ampullaris. Each crista is a transverse crest that projects inward into the lumen of the ampulla from the walls of the semi-circular duct of the otic labyrinth. The crista has a core of connective tissue and nerve fibers. The epithelium is of stereocilia-covered hair cells. The stereocilia are embedded in a thick extracellular gelatinous glycoprotein mass, the cupula. (200×)

FIGURE 21.36

LM of a cross section through the cartilaginous part of the eustachian (auditory) tube near the nasopharynx. The eustachian tube connects the tympanic cavity with the nasopharynx. The epithelial lining of the eustachian tube in the nasopharynx region is essentially pseudostratified cili-ated columnar epithelium with interspersed goblet cells. Elastic cartilage surrounds the eustachian tube in this area.

(200×)

Perichondrium — Lumen of eustachian tube

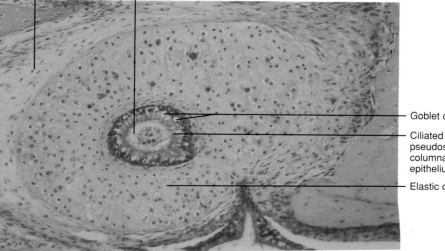

Goblet cells

Ciliated pseudostratified columnar epithelium

Elastic cartilage

Index

285